POLYMER SCIENCE AND TECHNOLOGY
Volume 16

POLYMERIC SEPARATION MEDIA

POLYMER SCIENCE AND TECHNOLOGY

A Continuation Order Plan is available for this series. A continuation order will bring delivery of each new volume immediately upon publication. Volumes are billed only upon actual shipment. For further information please contact the publisher.

POLYMER SCIENCE AND TECHNOLOGY
Volume 16

POLYMERIC SEPARATION MEDIA

Edited by

Anthony R. Cooper

Lockheed Missiles and Space Company, Inc.
Palo Alto Research Laboratory
Palo Alto, California

PLENUM PRESS • NEW YORK AND LONDON

Library of Congress Cataloging in Publication Data

Main entry under title:

Polymeric separation media.

   (Polymer science and technology; v. 16)
   "Collection of manuscripts based on presentations at a symposium...organized
for the second Chemical Congress of the North American Continent held in Las
Vegas, August 24–29, 1980."
   Includes bibliographical references and index.
   1. Membranes (Technology)–Congresses. 2. Polymers and polymerization–
Congresses. I. Cooper, Anthony R. II. Chemical Congress of the North American
Continent (2nd: 1980: Las Vegas, Nev.) III. Series.
TP159.M4P64   660.2'8423   82-3668

ISBN-13: 978-1-4613-3373-9   e-ISBN-13: 978-1-4613-3371-5
DOI: 10.1007/978-1-4613-3371-5

                                                  AACR2

Based on the proceedings of a symposium on Polymeric Separation Media,
organized for the Second Chemical Congress of the North American Continent,
and held August 24 – 29, 1980, in Las Vegas, Nevada

© 1982 Plenum Press, New York
Softcover reprint of the hardcover 1st edition 1982
A Division of Plenum Publishing Corporation
233 Spring Street, New York, N.Y. 10013

PREFACE

This volume is a collection of manuscripts based on presentations at a symposium "Polymeric Separation Media" organized for the Second Chemical Congress of the North American Continent held in Las Vegas, August 24-29, 1980.

The symposium was organized to bring together researchers in the expanding field of separations based on polymeric media. A diverse cross-section of research areas were presented, which were linked by the active separation agent being a polymeric material.

I would like to thank the authors for their endeavours and the audience for their participation, especially in light of the late change of venue for this meeting. Finally I am indebted to the Division of Polymer Chemistry Inc. of the American Chemical Society for their sponsorship.

Anthony R. Cooper, Ph.D., FRSC
Palo Alto, California
July 1981

CONTENTS

PART II

FUNCTIONALIZED POLYMERS AS SEPARATION MEDIA

PART III

POLYMERIC MEMBRANES AS SEPARATION MEDIA

DEDICATION

B.H.J. Hofstee (B.S., Doctorandus, Ph.D., all from the University of Groningen, The Netherlands) was Chief of the Division of Biochemistry, Palo Alto Medical Research Foundation. During his 30 year tenure at the Foundation, Barend acquired an international reputation for fundamental studies on esterases. He is best known for the "Hofstee plot" of second-order reactions and for the invention and development of hydrophobic chromatography of macromolecules.

Barend was preparing a manuscript for this symposium at the time of his death. In view of his contributions to modern science and his earned reputation as a gentleman of quality, we dedicate this volume to his memory.

Barend H.J. Hofstee 1912 - 1980

PUBLICATIONS OF BAREND H.J. HOFSTEE

T.P. Singer and B.H.J. Hofstee, Studies on wheat germ lipase. 1. Methods of estimation, purification and general properties of the enzyme, Arch. Biochem. 18, 229 (1948).

T.P. Singer and B.H.J. Hofstee, Studies on wheat germ lipase. II. Kinetics, Arch. Biochem. 18, 245 (1948).

B.H.J. Hofstee, The activation of urease, J. Gen. Physiol. 32, 339, (1949).

B.H.J. Hofstee, Inhibition of the enzymic oxidation of xanthine and xanthopterin by pteridines, J. Biol. Chem. 179, 633 (1949).

R.F. Schilling, J.S. Fruton, B.H.J. Hofstee, A.D. Welch, J.W. Harris, F.H. Gardner and W.B. Castle, Observations on the etiologic relationship of achylia gastria to pernicious anemia, XII. Failure of thymus aminopolypeptidase to act as intrinsic factor. J. Lab. and Clin. Med. 36, 942 (1950).

B.H.J. Hofstee, Spectrophotometric determinations of esterases, Science 114, 128 (1951).

B.H.J. Hofstee, On the evaluation of the constants V_m and K_m in enzyme reactions, Science 116, 329 (1952).

B.H.J. Hofstee, Specificity of esterases, I. Identification of two pancreatic aliesterases, J. Biol. Chem. 199, 357 (1952).

B.H.J. Hofstee, Specificity of esterases, II. Behavior of pancreatic esterases I and II towards a homologous series of n-fatty acid esters, J. Biol. Chem. 199, 365 (1952).

B.H.J. Hofstee, Specificity of esterases, III. Critical concentrations of the substrate, J. Biol. Che., 207, 211 (1954).

B.H.J. Hofstee, Specificity of esterases, IV. Behavior of horse liver esterase towards a homologous series of n-fatty acid esters, J. Biol. Chem., 207, 219 (1954).

B.H.J. Hofstee, Direct and continuous spectrophotometric assay of phosphomonoesterases, Arch. Biochem. Biophys. 51, 139 (1954).

B.H.J. Hofstee, On the mechanism of inhibition of xanthine oxidase by the substrate, J. Biol. Chem. 216, 235 (1955).

B.H.J. Hofstee, Alkaline phosphatase: I. Mechanism of action of Zn, Mg, glycine, Versene and hydrogen ions, Arch. Biochem. Biophys 59, 352 (1955).

B.H.J. Hofstee, Direct and continuous spectrophotometric assay of β-Blucosidase, Arch. Biochem. Biophys. 59, 398 (1955).

B.H.J. Hofstee, Graphical analysis of single enzyme systems, Enzymologia, 17, 273 (1956).

B.H.J. Hofstee, Fatty acid esters as substrates for trypsin and chymotrypsin, Biochim. Biophys. Acta, 24, 211 (1957).

B.H.J. Hofstee, Micelle formation in substrates of esterases, Arch. Bioch. Biophys. 78, 188 (1958).

B.H.J. Hofstee, A homologous series of soluble n-fatty acid esters as substrates for serum cholinesterase, J. Pharm. Exp. Ther., 123, 108 (1958).

B.H.J. Hofstee, Kinetics of β-glucosidase on ·the basis of intermediate enzymeglucoside formation, J. Am. Chem. Soc., 80, 3966 (1958).

B.H.J. Hofstee, Kinetics of chymotrypsin with a homologous series of n-fatty acid esters as substrates, Biochim. Biophys. Acta, 32, 182 (1959).

B.H.J. Hofstee, Non-inverted versus inverted plots in enzyme kinetics, Nature, 184, 1296, (1959).

B.H.J. Hofstee, Soluble stoichiometric complexes of DNA with chymotrypsin and trypsin, Biochim. Biophys. Acta, 44, 194 (1960).

B.H.J. Hofstee, Mechanism of action of bivalent metals and of phenothiazines on serum cholinesterase, J. Pharm. Exp. Ther., 128, 299 (1960).

B.H.J. Hofstee, Fatty acid esterases of low eserine sensitivity and related enzymes, Chapter 29, in "The Enzymes" (Boyer, Lardy and Myrback, editors) Academic Press, New York (1960).

B.H.J. Hofstee, Salt reversible inhibition of chymotrypsin by serum albumin and other proteins, J.Am. Chem. Soc., 82, 5166 (1960).

B.H.J. Hofstee, Composition and molecular shape of chymotrypsin-nucleic acid complexes, Biochem. Biophys. Res. (Comm. 4, 5 (1961).

B.H.J. Hofstee, Soluble complexes of nucleic acids with α-cyhmotrypsin and its derivatives, Biochim. Biophys. Acta, 55, 440 (1962).

B.H.J. Hofstee, Inhomogeneity of the chymotrypsins with respect to deoxyribonucleic acid binding, J. Biol. Chem., 238, 3235 (1963).

B.H.J. Hofstee, Some aspects of the specificity of fatty acid esterases, J. Histochem. Cytochem., 12, (9), 700 (1964).

B.H.J. Hofstee, Solubility and composition of protein-deoxyribonucleic acid complexes, Bioch. Biophys. Acta, 91, 340 (1964).

B.H.J. Hofstee, The rate of chymotrypsin autolysis, Arch. Bioch. Biophys., 112, 224 (1965).

B.H.J. Hofstee, Substrate hydrophobic groups and the maximal rate of enzyme reactions, Nature 213, (No. 5071), 42 (1967).

B.H.J. Hofstee, Renaturation of chymotrypsinogen A heated in 8M urea, Arch. Bioch. Biophys. 122, 574 (1967).

B.H.J. Hofstee, Crystallization and activation of chymotrypsinogen "X", Arch. Biochem. Biophys. 125, 1031 (1968).

B.H.J. Hofstee, Activation of chymotrypsinogen "X" and specificity of the enzyme, J. Biol. Chem., 243, 6306 (1968).

B.H.J. Hofstee, and D. Bobb, Heat denaturation of chymotrypsinogen A in the presence of polyanions, Biochim. Biophys. Acta 168, 565 (1968).

B.H.J. Hofstee, Effects of neutralization of amino groups on the kinetics of chymotrypsin, Bioch. Biophys. Res. Comm. 41, 1141 (1970).

D. Bobb and B.H.J. Hofstee, Gel isoelectric focusing for following the successive carbamylations of amino groups in chymotrypsinogen A, Anal. Bioch. 40, 209 (1971).

B.H.J. Hofstee, On the substrate activation of liver esterase, Bioch. Biophys. Acta, 258, 446 (1972).

B.H.J. Hofstee, Hydrophobic affinity chromatography of proteins, Analyt. Bioch., 52, 430 (1973).

B.H.J. Hofstee, Protein binding by agarose carrying hydrophobic groups in conjuction with charges, Bioch. Biophys. Res. Comm., 50, 751 (1973).

B.H.J. Hofstee, Immobilization of enzymes through non-covalent binding to substituted agaroses, Bioch. Biophys. Res. Comm. 53, 1137 (1973).

B.H.J. Hofstee, Hydrophobic aspects of protein binding by substituted agaroses, Polymer Preprints 15(1), 311, 1974.

B.H.J. Hofstee, Non-Specific Binding of Proteins by Substituted Agaroses, in: Immobilized Biochemicals and Affinity Chromatography (R.B. Dunlap, ed.) Plenum, New York, 1974, pp. 43-59.

B.H.J. Hofstee, Accessible Hydrophobic Groups of Native Proteins, Bioch. Biophys. Res. Comm., 63, 618-624 (1975).

B.H.J. Hofstee, Fractionation of Protein Mixtures through Differential Adsorption on a Gradient of Substituted Agaroses of Increasing Hydrophobicity, Prepar. Bioch. 5, 7-19 (1975).

B.H.J. Hofstee, Hydrophobic Effects in Adsorptive Protein Immobilization, J. Macromo. Sci (Chem) A10, 111-147 (1976). Also to be published as a chapter in "Polymer Grafts in Biochemistry" (H.F. Hixson, ed.), Dekker.

B.H.J. Hofstee, Hydrophobic Adsorption Chromatography, in "Methods of Protein Separation" (N. Catsimpoolas, ed.), Plenum, 1976, pp. 245-278.

B.H.J. Hofstee, Salt-Stable Hydrophobic versus Salt-Reversible Electrostatic Effects in Adsorptive Protein Binding, in: Enzyme Engineering, Vol. 3 (E.K. Pye, ed.) Plenum, New York, 1976, pp.347-355.

B.H.J. Hofstee, and N. Frank Otillio, Non-ionic Adsorption Chromatography of Proteins, J. of Chromat. 159, 57 (1978).

B.H.J. Hofstee, and N. Frank Otillio, Modifying Factors in Hydrophobic Protein Binding by Substituted Agaroses, J. of Chromat. 161, 153 (1978).

B.H.J. Hofstee, Immobilization of Enzymes through Hydrophobic and Other Non-Ionic Adsorption, FEBS, 52, 469 (1979).

B.H.J. Hofstee, Non-Ionic Adsorption Chromatography and Adsorptive Immobilization of Proteins, Pure and App. Chem. (IUPAC) 51, 1537 (1979).

B.H.J. Hofstee, Selective Aromatic-Hydrophobic Binding and Fractionation of Immunoglobulin by Means of Phenyl-$(CH_2)_n$-NH-substituted agaroses, Bioch. Biophys. Res. Comm., 91, 312 (1979).

B.H.J. Hofstee, Non-ionic Effects in Chromatographic Separation of Proteins through Differeitial Adsorptive Immobilization, Inserm. 86, 233 (1979).

SOME CONSIDERATIONS REGARDING THE

HYPERFILTRATION OF ORGANIC LIQUIDS

I.J. Brass and P. Meares

Chemistry Department
University of Aberdeen
Scotland

INTRODUCTION

The uses of polymeric membranes in the separation of gaseous mixtures and in the removal of solutes from aqueous solutions by hyperfiltration and electrodialysis are now well established if not yet widely applied. Less attention has been paid to the separation of organic mixtures by polymeric membranes. Gaseous hydrocarbons can be treated by the same methods as other gases but the selectivities of membranes giving fluxes large enough to be of serious interest are not sufficiently high to enable membrane processes to displace established methods of separation.

The separation of organic binary liquid mixtures by membranes can be attempted in two ways: pervaporation and hyperfiltration. Each process has attractive and unattractive features. Pervaporation, in which the liquid mixture is fed to one side of the membrane and the product as a vapour is withdrawn by pumping at the other side is practicable only for relatively volatile components. Under the restriction of idealized assumptions about the thermodynamic and kinetic properties of polymer + liquid mixtures, pervaporation should provide as good a separation factor and as large a flux as hyperfiltration but without the engineering complications associated with a feed stream at high pressure. There is experimental evidence however that hyperfiltration through hard polymer films can give larger fluxes than can pervaporation.[1]

This paper is devoted to a discussion of hyperfiltration because the technology of hyperfiltration plants is now well-developed. In any liquid separation process the choice of membrane

polymer presents crucial problems because the membrane must resist
dissolution or excessive swelling in the liquids, have a good
permeability towards one of them and show a high selectivity.   The
most promising developments so far have involved structural,
especially graft, copolymers[2] and highly polar polymers.[3]

THEORY

A comprehensive discussion of the wide range of phenomena that
can be anticipated in the hyperfiltration of non-ionic binary
mixtures was given by Schlögl[4] before the topic had acquired
technological interest and there were any data available.   Here we
begin with a very simplified treatment of steady state
hyperfiltration in order to explore the effects of some of the main
system parameters and variables and to provide a basis for
evaluating experimental data.

Consider two components, denoted by subscripts 1 and 2, whose
molar   volumes V_1 and V_2 are assumed to be constants at constant
temperature;   i.e. they are incompressible, and they mix with one
another and with the membrane polymer without volume change.   They
are assumed also to form ideal binary mixtures at all mole fractions
x_1 and x_2.

The equilibrium sorption of each liquid by the polymer is
assumed to be related to its mole fraction in the contacting
liquid mixture by a constant partition coefficient s_1 or s_2.   Thus
the concentrations $c_1(0)$ and $c_2(0)$ in the polymer in equilibrium
with a liquid phase of composition $x_1(0)$ and $x_2(0)$ are given by

$$c_1(0) = s_1 x_1(0) \tag{1}$$

$$c_2(0) = s_2 x_2(0) \tag{2}$$

If a homogeneous membrane of thickness ℓ is considered, the
quantities referring to the feed, high pressure, side are denoted
by the suffix (0) and those referring to the product side by
suffix (ℓ).   Experience so far is that a single pass through a
membrane cannot approach complete separation of the components.
More usually $x_1(0)$ and $x_1(\ell)$ do not differ as widely as one might
wish and the assumption of constant s_1 and s_2 over the composition
range covered in any particular experiment is not as improbable as
might, at first sight, appear.

Likewise we shall assume that each component has an activity
coefficient f and diffusion coefficient D in the membrane that are
independent of the concentrations of both components over the range
across the membrane.   Pressure is denoted by p, and R and T have
their usual meanings.

The flux density J_1 of component 1 across the plane distant x through the membrane is given by

$$J_1 = -\frac{D_1}{RT} c_1 \frac{d\mu_1}{dx} \tag{3}$$

where μ_1 is the chemical potential and c_1 the concentration at plane x. Within the assumptions listed

$$\mu_1 = \mu_1^{\,o} + RT \ln c_1 f_1 + p V_1 \tag{4}$$

whence

$$\frac{d\mu_1}{dx} = RT \frac{d \ln c_1}{dx} + V_1 \frac{dp}{dx} \tag{5}$$

In equation (3), D_1/RT represents the absolute mobility of component 1 according to the Nernst-Einstein relation. Thus diffusion is being treated as a consequence of random molecular Brownian motion.

On substituting equations (1) and (5) into (3) one finds

$$J_1 = -D_1 s_1 x_1 \left[\frac{d \ln x_1}{dx} + \frac{V_1}{RT} \frac{dp}{dx} \right] \tag{6}$$

The integration of this flux equation across the membrane requires a knowledge of the pressure profile. We shall assume it to be linear, so that

$$dp/dx = [p(\ell) - p(0)]/\ell = \Delta p/\ell, \text{ for all x} \tag{7}$$

Note, that for positive flux from 0 to ℓ, Δp is numerically negative.

Introducing equation (7) into (6) and integrating from x = 0, $x_1 = x_1(0)$ to x = ℓ, $x_1 = x_1(\ell)$ gives

$$J_1 = \frac{D_1 s_1}{\ell} \frac{\theta_1}{[1 - \exp(-\theta_1)]} [x_1(0) \exp(-\theta_1) - x_1(\ell)] \tag{8}$$

where θ_1 represents $V_1 \Delta p/RT$ (i.e. θ_1 is negative).

Paul[5] has proposed that instead of considering a pressure gradient in a supported membrane, the polymer should be regarded as uniformly at the higher pressure. Thus there would be a fall in concentration of sorbed liquid at the interface where polymer at the high pressure is in contact with liquid at low pressure. The concentration gradient thus produced within the membrane is supposed to give rise to normal Fickian diffusion.

When Paul's ideas are applied under the assumptions above there results

$$J_1 = \frac{D_1 s_1}{\ell} \frac{1}{\exp(-\theta_1)} [x_1(0) \exp(-\theta_1) - x_1(\ell)] \tag{9}$$

The second terms in the products on the right sides of equations (8) and (9) differ.   In practice $-\theta_1$ is likely to lie between 0.1 ($\sim$ 20 atm) and 2.0 ($\sim$ 400 atm).   Equation (8) predicts a larger flux than equation (9).   The difference is about 5% when $-\theta_1 = 0.1$, 27% when $-\theta_1 = 0.5$ and 100% when $-\theta_1 = 1.6$.

Where tests have been possible[6,7] equation (9) has been found to predict a flux rather less than that observed.   This might be a reason for preferring equation (8).   Since we know neither s_1 nor D_1 and shall attempt to predict only the behaviour of separation factors (ratios of fluxes) it makes no difference which equation is used.   The calculations given here have been based on equation (8).

An equation analogous to (8) but with altered subscripts gives the flux density of component 2.   Furthermore in hyperfiltration the product composition is determined by the fluxes.   Hence

$$J_1/J_2 = x_1(\ell)/x_2(\ell) \tag{10}$$

On dividing equation (8) by its counterpart for component 2 and introducing equation (10) one obtains

$$\frac{x_1(\ell)}{x_2(\ell)} = \gamma_{12} \frac{V_1}{V_2} \frac{[x_1(0) \exp(-\theta_1) - x_1(\ell)][1 - \exp(-\theta_2)]}{[x_2(0) \exp(-\theta_2) - x_2(\ell)][1 - \exp(-\theta_1)]} \tag{11}$$

where the membrane selectivity coefficient γ_{12} is defined by

$$\gamma_{12} = D_1 s_1/D_2 s_2 \tag{12}$$

For a given feed composition $x_1(0)$, $x_2(0)$, and operating pressure $-\Delta p$, equation (11) can be solved for the product composition $x_1(\ell)$ since $[x_1(\ell) + x_2(\ell)] = 1$.   If the system parameters V_1, V_2 and γ_{12} are known numerical calculations of $x_1(\ell)$ can be performed.

PREDICTED IDEALIZED BEHAVIOUR

The treatment given above makes no allowance for coupling of flows and consequently can predict only congruent fluxes i.e. both components moving down their thermodynamic potential gradients.

The effectiveness of a separation can conveniently be represented in two ways.   Either the change of composition Δx_1

brought about by a single pass through the membrane, where $\Delta x_1 = [x_1(\ell) - x_1(0)]$, can be plotted as a function of the feed composition $x_1(0)$ and the operating pressure; or the behaviour of the separation factor β_{12}, defined by

$$\beta_{12} = \frac{x_1(\ell)/x_1(0)}{x_2(\ell)/x_2(0)} \tag{13}$$

can be examined.

In carrying out some calculations to illustrate the predicted behaviour of mixtures undergoing hyperfiltration we have selected component 1 as that to which the membrane is selectively permeable. Thus γ_{12} and β_{12} are always greater than unity.

Molar volumes of 100 mℓ and 200 mℓ have been considered for illustration. If smaller molar volumes had been chosen larger pressures would have been needed to achieve the same degrees of separation. All calculations refer to 298K.

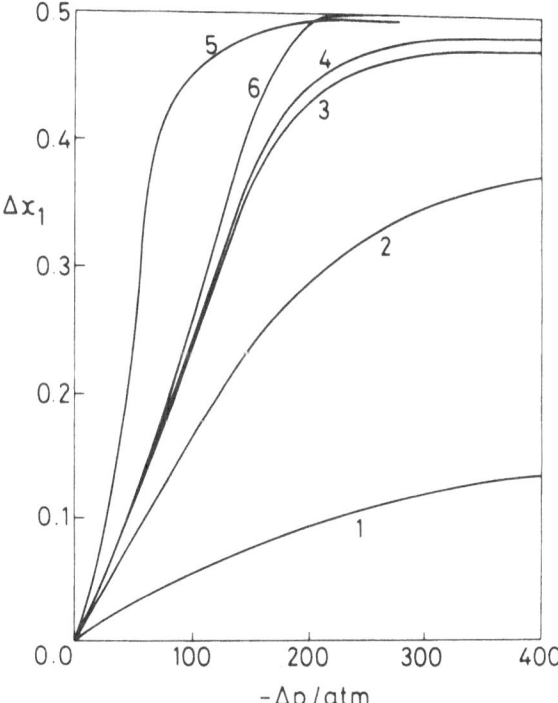

Fig. 1. The effect of pressure on the separation Δx_1 of a mixture in which $x_1(0) = x_2(0) = 0.5$. (1) $V_1 = V_2 = 100$ mℓ, $\gamma_{12} = 2$; (2) $V_1 = V_2 = 100$ mℓ, $\gamma_{12} = 10$; (3) $V_1 = 100$ mℓ, $V_2 = 200$ mℓ, $\gamma_{12} = 100$; (4) $V_1 = V_2 = 100$ mℓ, $\gamma_{12} = 100$; (5) $V_1 = 200$ mℓ, $V_2 = 100$ mℓ, $\gamma_{12} = 100$; (6) $V_1 = V_2 = 100$ mℓ, $\gamma_{12} = 1000$.

Figure 1 shows the effect of the operating pressure on Δx_1 at several values of γ_{12} with $V_1 = V_2$ and for $V_1/V_2 = 0.5$, 1.0 and 2.0 for $\gamma_{12} = 100$. It is clear that for $V_1/V_2 \leqslant 1$ there is a marked improvement in separation with increasing pressure up to about 200 atm. At 100 atm there is only a slight advantage in making γ_{12} greater than 100. In Figure 1 the feed composition has been held constant at $x_1(0) = x_2(0) = 0.5$, thus the maximum imaginable value of Δx_1 is 0.5.

When $V_2 > V_1$ the pressure makes a bigger contribution to the force on 2 than on 1. Nevertheless curves 3 and 4 scarcely diverge below 150 atm and the penalty of having an unfavourable molar volume ratio (preferred component having the lower molar volume) is not large. By contrast, when $V_1 > V_2$ (curve 5) there is a great advantage. Not only is the ultimate performance at high pressure significantly improved but, more importantly for practical application, an efficient separation is achieved at a much lower pressure.

Figure 2 illustrates the effect of feed composition on the separation factor in several cases. β_{12} has an upper limit of $\gamma_{12} V_1/V_2$. This is approached as the pressure and $x_1(0)$ are increased. In all cases separation is poor when $x_1(0)$ is small but improves sharply once $x_1(0)$ exceeds a critical value which depends on $-\Delta p$ and V_1/V_2. Figure 2 suggests that hyperfiltration may prove to be most useful for eliminating substantial amounts of impurity from the more permeable component but would be less effective for recovering a permeating component present in only small amount.

Here we have treated only steady states. Small amounts of scarcely permeating components may be recoverable from the retentate in batch hyperfiltration i.e. under non-steady flow conditions. Diagrams based on Schlögl's m-representation[4] should be useful for exploring the circumstances under which such procedures would be effective.

One particular application of the hyperfiltration of solvent mixtures that has been suggested is the separation of azeotropes.[3] In such examples Δx_1 is of particular interest. Figure 3 shows some calculations carried out with $\gamma_{12} = 100$. Increasing γ_{12} to 1000 has only a very small effect on the curves.

With $V_1 = V_2 = 100$ ml (curves 2, 3, 5) increasing $-\Delta p$ above 100 atm not only increases Δx_1 at all compositions, it also shifts the maximum in Δx_1 further towards low $x_1(0)$. Of course the curves can never lie above the -45° slope drawn through $x_1(0) = 1$. This explains why all the graphs in Figure 3 coincide on the right side of the diagram. Setting $V_1 < V_2$ (curve 4) at 200 atm has

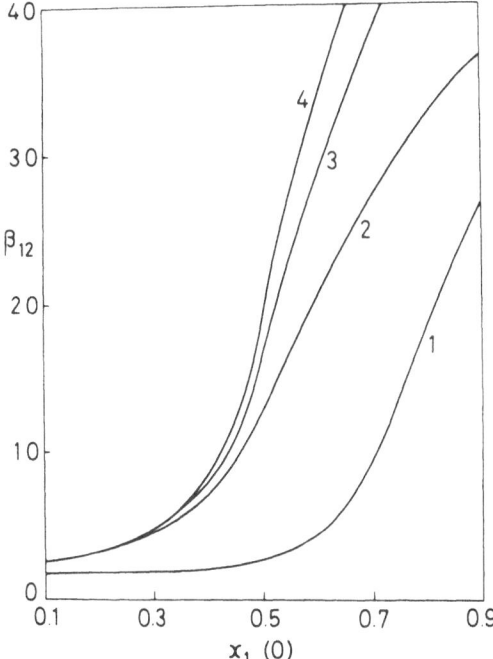

Fig. 2. Effect of feed composition on the separation factor β_{12}
for selectivity $\gamma_{12} = 100$.   (1) $V_1 = V_2 = 100$ ml, $-\Delta p = 100$
atm;   (2) $V_1 = 100$ ml, $V_2 = 200$ ml, $-\Delta p = 200$ atm;   (3) $V_1 =$
$V_2 = 100$ ml, $-\Delta p = 200$ atm;   (4) $V_1 = 200$ ml, $V_2 = 100$ ml,
$-\Delta p = 100$ atm.

very little effect but when $V_1 > V_2$ (curve 1) an excellent
separation is achievable under 200 atm pressure at all values of
$x_1(0)$ above about 0.25.   Naturally azeotropes cannot realistically
be regarded as ideal mixtures.

COMPARISON WITH EXPERIMENTAL DATA

        The corresponding Figure 3 in the preprinted abstract of this
paper purported to compare the observed data of Franck and
Hungerhoff[3] on methanol and carbon tetrachloride hyperfiltered at
60 atm through poly-acrylonitrile with a curve calculated for $\gamma_{12} =$
125.   The former was drawn correctly;   the latter was calculated
wrongly and is completely in error.[*]   While carrying out the

* The same error appears in ref. 8, Figure 1

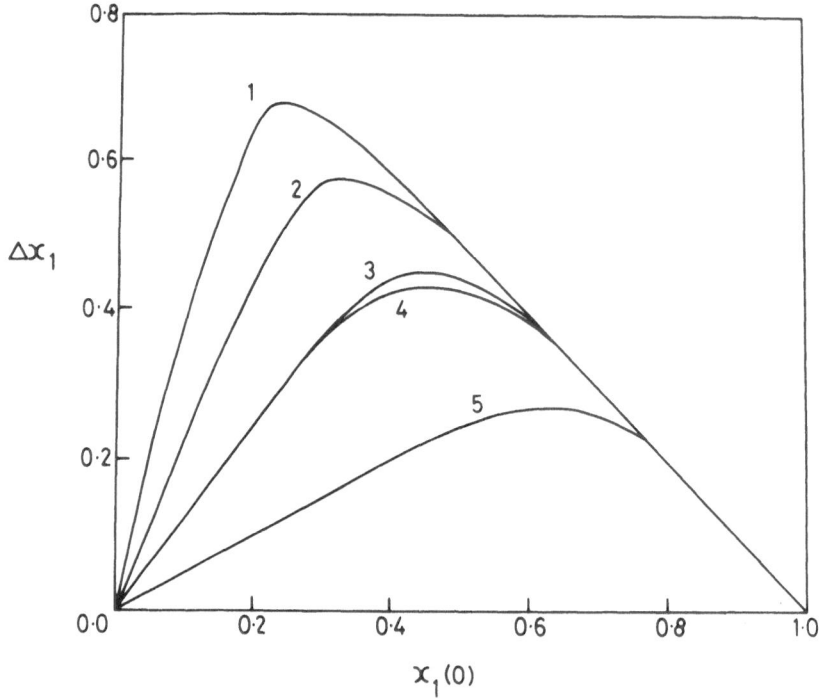

Fig. 3. Effect of feed composition on the separation Δx_1 for $\gamma_{12} =$
100.   (1) $V_1 = 200$ mℓ, $V_2 = 100$ mℓ, $-\Delta p = 200$ atm; (2) $V_1 =$
$V_2 = 100$ mℓ. $-\Delta p = 300$ atm;   (3) $V_1 = V_2 = 100$ mℓ, $-\Delta p = 200$ atm;
(4) $V_1 = 100$ mℓ, $V_2 = 200$ mℓ, $-\Delta p = 200$ atm;   (5) $V_1 = V_2 =$
100 mℓ, $-\Delta p = 100$ atm.

calculation the molar volumes of methanol, 40.7 mℓ, the carbon
tetrachloride, 97.8 mℓ, were inadvertently interchanged.

    When one attempts to carry out the calculation correctly it is
found that the quantity in square brackets in equation (8) is
negative for methanol, the preferentially permeating component,
when calculated from the observed feed and product compositions
at values of $x_1(0)$ less than about 0.55.   This would mean that
methanol was transported against its thermodynamic gradient, i.e.
incongruently, at mole fractions below 0.55.   Although such a
result does not contradict the laws of thermodynamics, provided the
combined transport process of the two components results in an
overall dissipation of free energy, it could only occur if there
were a strong coupling between the flows of the two components, an
effect specifically excluded from the idealized theory.

    Strong flow coupling and, especially incongruent flow, in
binary mixtures is rare.[4]   To assess the situation properly in the
present case it is essential to replace mole fractions by activities
a_1 and a_2.   The condition for congruent flow of the preferred

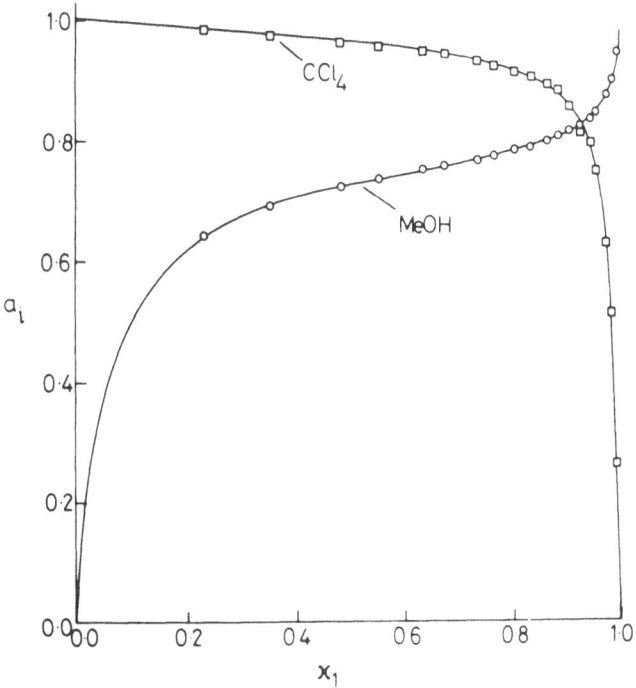

Fig. 4.   Plots of activities a_1 and a_2 versus mole fraction of
          methanol x_1 for methanol + carbon tetrachloride at 20°C.[9]

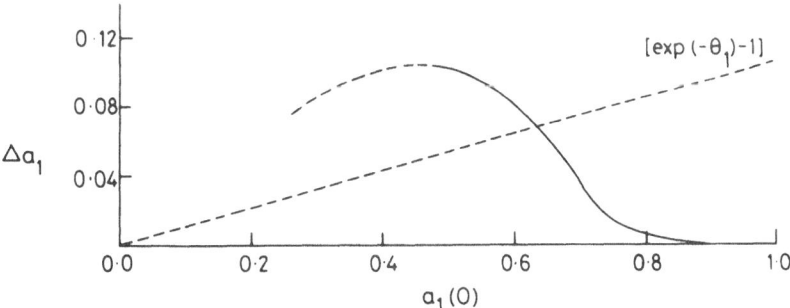

Fig. 5. Δa_1 versus activity of methanol $a_1(0)$ in the feed for
          hyperfiltration of methanol + carbon tetrachloride at
          60 atm.   (Based on data from ref. 3 and 9).   The dashed
          line has a slope of $[\exp(- V_1 \Delta p/RT) - 1]$.   Where the curve
          lies above this line the transport of methanol is
          incongruent.

component is then

$$\Delta a_1/a_1(0) < [\exp(-\theta_1) - 1] \tag{14}$$

The flow of component 2 is certain to be congruent because it travels down its concentration and pressure gradients.

Figure 4 shows the plots of activity versus mole fraction for methanol and carbon tetrachloride in their mixtures estimated from partial vapour pressure data.[9]   The system shows large positive deviations from ideality and over a considerable range of composition the activities scarcely vary.

Figure 5 shows a plot of Δa_1 versus $a_1(0)$ constructed from the experimental separation data[3] at 60 atm and Figure 4.   Also shown on Figure 5 is a line of slope $[\exp(-\theta_1) - 1]$ drawn through the origin.   For flow to be congruent the Δa_1 curve must lie below this line.   It appears that at low concentrations of methanol its flow is almost certainly incongruent but we cannot claim a high degree of accuracy for our measurements of the published hyperfiltration curves and cannot therefore be dogmatic about our findings as yet.

CONCLUSION

Even if the observed Δa_1 curves only approach very closely, without actually crossing, the limiting line it strongly suggests that coupling between the flows of the two components is occurring to a substantial degree.   Coupling will therefore have to be included in any more realistic theory of hyperfiltration.

Furthermore, in order to make systematic rather than purely empirical progress in the development of hyperfiltration for the separation of organic liquid mixtures the coupling or, preferably, molecular friction coefficients must be measured for selected mixtures and membranes.   An analysis of this problem in terms of non-equilibrium thermodynamics, including an identification of the necessary experimental measurements, was given several years ago by one of the authors.[10]   To date no experimental work on this scheme has been published and it is hoped that this omission may be remedied soon.

REFERENCES

1. J.B. Craig, P. Meares and J. Webster, in Diffusion Processes,
    J.N. Sherwood et al., eds., Vol. 1, p.609, Gordon & Breach,
    London (1971).
2. P. Aptel, N. Challard, J. Curry and J. Neel, J. Membrane Sci.,
    1: 271 (1976).

3. U.F. Franck and J. Hungerhoff, Internat. Scand. Congr. Chem.
    Eng. (Proc.) Symp. IV: 1 (1974).
4. R. Schļögl, Stofftransport durch Membranen, Ch. 3, Fortshr.
    physik. Chem., 9: Steinkopff Verlag, Darmstadt (1964).
5. D.R. Paul, J. Appl. Polym. Sci., 16: 771 (1972).
6. D.R. Paul and J.D. Paciotti, J. Polym. Sci., Polym. Phys.,
    13: 1201 (1975).
7. F.W. Greenlaw, W.D. Price, R.A. Shelden and E.V. Thompson,
    J. Membrane Sci., 2: 141 (1977).
8. P. Meares., Ber. Bunsenges. physik. Chem., 83: 342 (1979).
9. J. Timmermans, The Physico-chemical Constants of Binary Systems
    in Concentrated Solutions, Interscience Publishers Ltd., New York,
    2: 277 (1959).
10. P. Meares, Pure & Appl. Chem., 39: 99 (1974).

MEASUREMENT OF TRANSPORT INTERACTION IN MEMBRANES

K. S. Spiegler, T. S. Brun and A. Berg

Department of Chemistry and Chemical Engineering
Michigan Technological University
Houghton 49931

To determine the minimum number of transport parameters for a given type of membrane system, precise measurements at different stationary states must be combined. Close control of the boundary conditions in the stationary states is therefore necessary.

The mechanical and transport characteristics of many membranes depend on the concentrations of the solutions separated by the membrane. Hence the "concentration clamp" method was devised to keep these concentrations constant during any transport experiment. (Zelman, Kwak and Spiegler, 1971; Schmitt and Spiegler, 1974) Conductivity cells are inserted in the two half-cells separated by the membrane. The resistances of the cell solutions are compared to that of a reference resistor by a very accurate impedance comparator. The output of this comparator, which can be positive, negative, or zero, is then fed into a phase-sensitive amplifier to open or close a relay. At the "depleted" (salt-donor) side, this relay activates an automatic buret which pushes a concentrated electrolyte solution into the system until the reference concentration is reached again. At the "enriched" (salt-acceptor) side of the membrane, the relay activates a pump which circulates the contents of a reservoir of known volume connected to the "enriched" half-cell. This reservoir is originally filled with deionized water, and its contents serve to dilute the solution in the "enriched" half-cell when needed. A small portion of the solution in the half-cell passes through the reservoir while most of the solution is recirculated without change. Pumping continues until the reservoir has released enough water to lower the concentration to that of a reference cell. The concentration of the solution in the reservoir is continuously being monitored by measurement of its electric conductivity. This system keeps the concentration at either side of the membrane constant within 0.02% even when an electric current is passed, since electrodes are chosen such as to create no pH changes nor inject foreign ions. The salt and water transfer can be calculated from the volume of the concentrated solution added by the automatic buret and the volume change in the cell compartments.

The (a.c.) electrical resistance of the membrane is measured in a cell of similar dimensions, as a function of the distance between the electrodes. Two platinized platinum electrodes of 1.90 inch diameter can be moved forwards and backwards in the cell by means of a screwdriven mechanism, such that the electrodes remain parallel to the plane of the membrane; displacements to one-thousandth of an inch can be monitored with two dial-indicators which are mounted on both sides of the cell. One electrode usually remains stationary close to the membrane, but not in direct contact with it. The other is moved along the cylinder axis. The membrane holder, the membrane support and the membrane itself are the same as used with the transport cell.

Osmosis - dialysis, electromigration - electroosmosis, and hydraulic permeation are measured in the "concentration clamp" cell. The determination of membrane resistivity is based on a substitution method, i.e., the resistance between the electrodes is measured (a) with the membrane in the holder and (b) with the membrane removed from the holder, but with a membrane ring on the periphery of the two perforated membrane supports.

Since the construction of the first concentration-clamp cell, important additional results were obtained with similar cells by Delmotte and Chanu; Halary, Noël and Monnerie; and Hurwitz.

REFERENCES

Delmotte, M. and J. Chanu, Electrochimica Acta 18, 963 (1973).
Halary, J. L., C. Noël and L. Monnerie, J. Appl. Polym. Sci., 24 985 (1979).
Halary, J. L., Relations entre la structure et les phenomènes de transport dans les membranes de diacétate de cellulose pour osmose inverse, Thèse de doctorat d'état ès sciences physiques, Université Pierre et Marie Curie, Paris 6, Dec. 18, 1979.
Hurwitz, H. D., private communication, Université Libre de Bruxelles, Belgium.
Schmitt, A. and K. S. Spiegler, Unpublished results (1974).
Zelman, A., J. C. T. Kwak, J. Leibovitz and K. S. Spiegler, Experientia Suppl. 18, 679 (1971).

FREE VOLUME ESTIMATES IN CRYSTALLINE AND FILLED POLYMERS

H. L. Frisch   A. E. Kreituss

Department of Chemistry   Institute of Wood Chemistry
SUNY at Albany   Academy of Sciences of the
Albany, NY 12222   Latvian SSR
                          Riga, Latvian SSR 226006, USSR

Introduction

Diffusion of low molecular weight penetrants in crystalline and filled polymer films has been extensively studied recently using sorption, permeation and micro-interferometric techniques in large temperature intervals both below and above the fusion temperature (1,2). Consistent with other observations diffusion appears to be restricted to the amorphous, non-filled regions of the polymer system. We attempt to correlate the behavior of diffusion coefficients D, obtained by measuring $t_{1/2}$ (3) in sorption with free volume changes caused by the presence of crystallinity or fillers, suggested by a Fujita-like treatment of the data.

The polymers investigated were polyethylene (PE) and poly-propylene (PP) with several degrees of crystallinity, Φ_c, as well as their random copolymers (EP) with propylene weight fraction from 0.01 to 0.7 and $0 \leq \Phi_c < 0.7$. The penetrants used were CCl_4, benzene and n-hexene. The fillers used were talc (specific surface area 6.1 m^2/g(B.E.T.), density 2.75g/cc, mean particle diameter 7μ and alumina (36.9 m^2/g(B.E.T.), density 3.97g/cc, mean particle diameter 12μ. The volume fraction of filler Φ_f, employed in both the amorphous and crystalline systems satisfied $0 \leq \Phi_f \leq 0.3$.

After introducing the simple free volume theory we compare it with the experimental data. The correlation allows one to obtain the rough degrees of crystallinity approximately from diffusion data and surprisingly good agreement is obtained between the calculated and experimental concentration dependence of the diffusion coefficients.

Theory

We write for the diffusion coefficient of a partially crystalline sample the free volume expression

$$\ln \frac{D}{D(\Phi_1=0)} = \frac{B}{f_2} - \frac{B}{f_{12}} \tag{1}$$

where f_2 is the free volume fraction of pure polymer and the free
volume fraction $f_{12}(\Phi_1,\Phi_c)$ depends on the volume fractions of
penetrant Φ_1 and degree of crystallinity Φ_c through the density of
the polymer sample $d(\Phi_1,\Phi_c)$, viz.

$$f_{12}(\Phi_1,\Phi_c) = \frac{v-v_0}{v} = \frac{d_0-d(\Phi_1,\Phi_c)}{d_0} \underset{\sim}{} \frac{d_c-d(\Phi_1,\Phi_c)}{d_c} \tag{2}$$

with $v = d^{-1}(\Phi_1,\Phi_c)$ the specific volume of the actual polymer
sample and $v_0 = d_0^{-1}$ the specific volume of the polymer at $0°K$
which we take to be the density of the 100% crystalline polymer d_c,
in first approximation.

The free volume fraction $f_{12}^{am}(\Phi_1)$ of the corresponding 100%
amorphous polymer $f_{12}^{am}(\Phi_1) = f_{12}(\Phi_1,\Phi_c=0)$ can be represented
according to Fujita et al. (4) as

$$f_{12}^{am}(\Phi_1) = f_2 + (f_1-f_2)\Phi_1 = f_2 + \beta\Phi_1 = \frac{d_c-d_a(\Phi_1)}{d_c} \tag{3}$$

where f_i is the free volume fraction of the pure penetrant i=1
and the amorphous polymer for i=2, respectively, and $d_a(\Phi_1) =$
$d(\Phi_1,\Phi_c=0)$.

Making the additivity assumption on the densities (which is
consistent with eq.(3) as used by Fujita),

$$d(\Phi_1,\Phi_c) = d_a(\Phi_1)(1-\Phi_c) + d_c\Phi_c \tag{4}$$

and substituting eq.(4) into eq.(2), and using eq.(3), one obtains

$$f_{12}(\Phi_1,\Phi_c) = [f_2 + \beta\Phi_1](1-\Phi_c) \tag{5a}$$

to be employed in eq.(1) resulting in

$$\ln \frac{D}{D(\Phi_1=0)}^{-1} = \frac{f_2(1-\Phi_c)}{B} + \frac{f_2^2(1-\Phi_c)}{\Phi_1 B\beta} \tag{5b}$$

The resulting diffusion coefficient of a partially crystalline
polymer reflects effects due to the presence of the crystalline
component on the transport properties of the amorphous phase (thus
the polymer sample is not treated as a medium composed of two
inert phases).

In the case of added filler of volume fraction Φ_f eq.(2) must be replaced by

$$f_{12}^{(f)}(\Phi_1,\Phi_c,\Phi_f) = \frac{d_c - d_f(\Phi_1,\Phi_c,\Phi_f)}{d_c} \qquad (6)$$

which together with the additivity assumption

$$d_f(\Phi_1,\Phi_c,\Phi_f) = d_f\Phi_f + d(\Phi_1,\Phi_c)(1-\Phi_f) \qquad (7)$$

where d_f if the pure filler density.

Eq.(7) is expected to break down at sufficiently large filler concentrations due to formation of voids on incompletely wetted filler particles. We will not correct here our simple theory for such effects, reserving this for a later publication.

Using eq.(7) in eq.(6) we obtain

$$f_{12}^{(f)}(\Phi_1,\Phi_c,\Phi_f) = f_{12}(\Phi_1,\Phi_c) - [\frac{d_f-d(\Phi_1,\Phi_c)}{d_c}]\Phi_f =$$
$$= f_{12}(\Phi_1,\Phi_c) - \beta_f\Phi_f \qquad (8)$$

to be used in eq.(1) instead of f_{12}.

Comparison with Experiments and Discussion

In Figure 1 are shown the experimental and theoretically calculated concentration dependence of D of CCl_4 in ethylene-propylene copolymers EP-x (where x designates weight percent of propylene in the copolymer). EP-40 is amorphous and parameters B and f_2 given in Figure 1 were obtained by fitting eq.(3) to the initial data to produce the uppermost curve in Figure 1. Using these parameters and the average value of the experimentally determined volume fraction of crystallinity Φ_cx-ray and $\Phi_c$density from x-ray and density measurements respectively in eq.(5), we obtained the full line curves for the partially crystalline EP-1($\bar{\Phi}_c$ = 0.68) and EP-10($\bar{\Phi}_c$ = 0.49). The agreement between theory and experiment is quite satisfactory.

Figure 2 shows the zero penetrant diffusion coefficient $D(\Phi_1=0)$ as a function of the propylene weight fraction in various EP-x copolymers for the three diluents. The solid lines in computed from theory, i.e. using eq.(5),

$$D_{EP-x} = E_{EP-40}(\Phi_1=0)\exp[-B\Phi_c/f_2(1-\Phi_c)] \qquad (9)$$

Plot (a) in Figure 3 shows the decrease of diffusivity with increasing crystallinity (experimental points) and is the basis

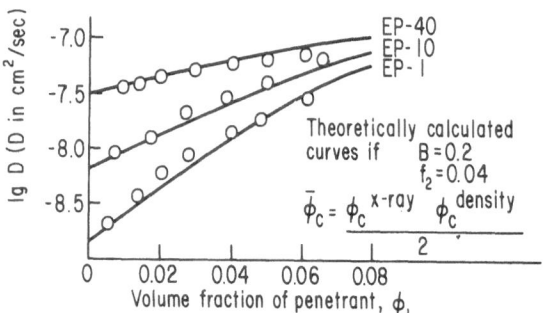

Figure 1.   Experimental and theoretically calculated concentration
            dependence of the diffusion coefficient, F of CCL_4 in
            amorphous (EP-40) and crystalline (EP-1, EP-10)
            ethylene-propylene copolymers at 25°C:

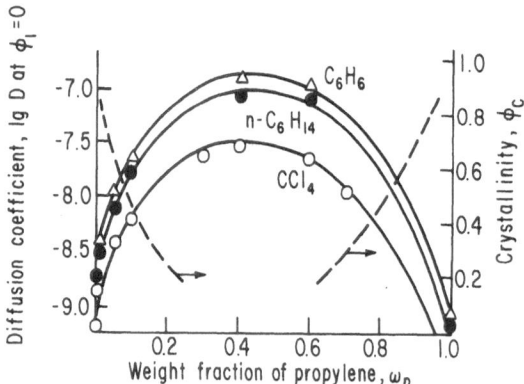

Figure 2.   Calculated (lines) and experimentally obtained varia-
            tion of the diffusion coefficient, $D(\phi_1=0)$ at zero
            concentration in ethylene-propylene copolymers and
            homopolymers at 25°C.   The broken lines - theoretically
            calculated degree of crystallinity.

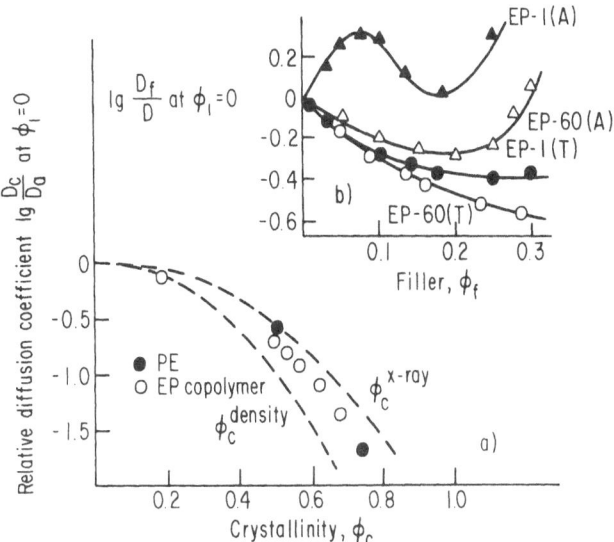

Figure 3.   Variation of the diffusion coefficient of CCl₄ with
            crystallinity (a) and n-hexane with filler content
            (b) in ethylene-propylene copolymers at 25°C. Filler
            A - alumina; T-talc; EP-1 and EP-60; 1 and 60 wt. %
            of propylene in copolymer.

for our contention that transport measurements can provide rough
estimates of crystallinity at least in this system. The dotted
curves show the computed decrease in diffusivity from Φ_c obtained
from density or x-ray measurements.

Figure 3(b) shows the more complex behavior of the same filled
amorphous (EP-60) and crystalline (EP-1) polymers. The concentration dependence of these filled polymers is not described as well
by eq.(8) using only experimentally determined Φ_c, Φ_f and parameters
B and f_2 from the amorphous polymers, eq.(7) requires modification
to take into account of void formation and the size of the filler
particles. Eq.(8) roughly describes the experimentally obtained
data providing experimental degree of crystallinity data is
employed in eq.(8), since crystallinity changes markedly with
filler concentration as can be seen from the Table 1 below:

TABLE 1

Crystallinity for Filled Crystalline Copolymers		
Polymer	Filler Content Φ_f, vol. %	Crystallinity Φ_c^{x-ray}, %
EP-1	0	77
EP-1 + talc	5	65
" "	25	30
EP-1 + alumina	5	35
" "	25	10

Acknowledgement

This work was supported by NSF DMR 7805938.

References

1. A. E. Kreituss, A. Ye. Chalykh and A. Ya. Metra, Modification
of Polymer Materials, Riga Polytechnical Institute Press,
Latvian SSR, 5, 217 (1975).
2. A. E. Kreituss et al., Modification of Polymer Materials, Riga
Polytechnical Institute Press, Latvian SSR, 7, 35 (1978).
3. J. Crank, The Mathematics of Diffusion, 2nd Ed., p. 238,
Clarendon Press, Oxford (1975).
4. H. Fujita, Advan. Polymer Sci. 3, 1 (1961).

DIFFUSION COEFFICIENTS OF POLYSTYRENES IN POROUS GELS FROM MASS
TRANSFER DISPERSION IN GEL PERMEATION CHROMATOGRAPHY

John V. Dawkins and Graham Yeadon*

Department of Chemistry
Loughborough University of Technology
Loughborough, Leicestershire LE11 3TU, England

ABSTRACT

For high polymers separating by gel permeation chromatography, experimental data for plate height for non-permeating polymers suggest that the contribution to chromatogram broadening from solute dispersion in the mobile phase is hardly influenced by eluent flow rate. Consequently, the overall expression for plate height for permeating polymers is simplified considerably, permitting the evaluation of the contribution to chromatogram broadening from solute mass transfer in the stationary phase. From the experimental dependence of plate height for polystyrene standards on eluent flow rate for microparticulate packings with particle diameters in the range 8-20 μm, the mass transfer contribution has been determined from which the diffusion coefficient of polystyrene in the pores in the stationary phase has been estimated. The values of diffusion coefficient are below theoretical expectation but are in agreement with the results of other experimental studies.

INTRODUCTION

The hindrance to solute diffusion in porous media due to tortuosity and constriction effects is important in many transport processes, e.g. in chromatography and membrane separations. In the first detailed theoretical treatment of chromatogram broadening in gel permeation chromatography (GPC), Giddings and Mallik[1] indicated from experimental evidence for soft gels[2] that the diffusion coefficient of a solute D_S in a porous gel was about $2D_m/3$ where D_m

---
*Present address: National Adhesives and Resins, Braunston, Daventry, Northants, England.

is the solute diffusion coefficient in the mobile phase. Values of
D_S may be determined by evaluating from the total plate height H of
an experimental chromatogram the contribution arising from solute
mass transfer in the stationary phase. Giddings et al.[3] studied
the chromatogram broadening of low molecular weight polystyrenes
separated by GPC with rigid porous glass particles having diameters
in the range 44-74 μm and found that D_S was about $D_m/6$. In this
paper, we report plate height data for GPC separations of polystyrene
(PS) with macroporous silica microspheres and show how the chromatogram broadening term arising from mass transfer in the stationary
phase may be evaluated by a simple procedure from which values of
D_S are determined.

EXPERIMENTAL

    The samples of laboratory-prepared macroporous silicas des-
ignated H4 and SG30 were kindly provided by Dr.J.D.F.Ramsay and Dr.
D.C.Sammon of AERE Harwell. Examination of the silica microspheres
by scanning electron microscopy, as described previously,[4] and
Coulter Counter suggested that the mean particle diameter was
16.6 μm (H4) and 8 μm (SG30). A silica designated γ-G SG30 was
prepared by reacting SG30 silica with γ-glycidoxypropyltrimethoxy-
silane in aqueous media according to the method described by
Regnier and Noel.[5] The H4 silica microspheres were packed into
stainless-steel columns (20 cm x 0.3 cm ID) using the technique
described by Majors.[6] Columns (25 cm x 0.8 cm ID) of SG30 and γ-G
SG30 silicas were packed by pumping a methanol slurry according to
the method of Bristow et al.[7] At the top of the column the silica
packing was covered by a stainless steel mesh above which was placed
Ballotini which in turn was covered by a porous Teflon plug. A
syringe injection head was then attached at the top of the column.

    High-performance GPC separations were performed with a Perkin-
Elmer Model 1220 positive displacement syringe pump (flow settings
0.05 - 6.00 cm^3 min^{-1}, < 3000 psi, 500 cm^3 capacity) and with an
Applied Research Laboratories Limited ultra-violet detector (254 nm,
cell volume = 8μdm^3). The eluent was tetrahydrofuran (BDH Chemicals
Ltd.) which was destabilized, distilled and degassed before use.
Column efficiencies over a range of eluent flow rates for polystyrene standards (Waters Associates Inc.) which are designated PS
followed by a number corresponding to the molecular weight were
established with a solute concentration of 0.2% w/v in injection
volumes of 2 μdm^3 (H4 silica) and 10 μdm^3 (SG30 and γ-G SG30
silicas). Plate height was calculated from the plate number determined from the width of an experimental chromatogram at half its
height. The linear velocity of the eluent for each column was
calculated with the interstitial (or void) volume V_o which was
estimated from the molecular weight calibration curve established
with polystyrene standards.

RESULTS AND DISCUSSION

Experimental data for plate height H for a solute having
constant retention volume V_R over the range of linear flow rate u
of the eluent may be interpreted in terms of the dispersion mechanisms occurring in the mobile and stationary phases. Giddings and
Mallik[1] proposed an expression for H for GPC separations, which for
a monodisperse solute we shall write in the form

$$H = (B/u) + C_s u + \Sigma \, 1/[(1/A) + (1/C_m u)] \qquad (1)$$

in which A, B, C_s and C_m are coefficients depending on several
parameters (see later) where the first term results from dispersion
owing to molecular diffusion in the longitudinal direction in the
mobile phase, the second term results from solute dispersion owing
to mass transfer in the stationary phase, and the third term
containing contributions from eddy diffusion (A) and mass transfer
($C_m u$) results from solute dispersion in the mobile phase. There is
abundant experimental evidence, as reviewed elsewhere,[8] indicating
that the first term in equation (1) may be neglected for high
polymers at practical flow rates, e.g. u > 1 mm s^{-1}. Experimental
plate height data plotted as a function of u in Fig. 1 do not display
the minimum required by the first term in equation (1). The polystyrene standard PS-1987000 in Fig. 1 may be regarded as a non-
permeating solute, see the GPC calibration curve for H4 silica

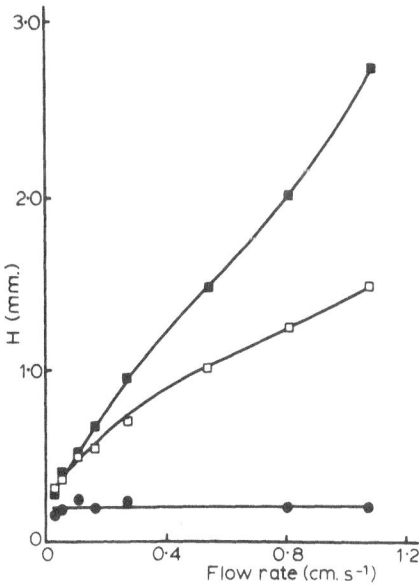

Fig. 1   Dependence of plate height on flow rate for H4 silica:
■, PS-9800; □, PS-35000; ●, PS-1987000.

reported in an earlier study.[4] The plate height data for this
excluded solute shown in Fig. 1 suggest that for high polymers
chromatogram broadening due to solute dispersion in the mobile phase
exhibits little or no change with the eluent flow rate u, in agreement with experimental data for non-permeating polymers reviewed
elsewhere.[8] It may be shown that this is the expected behaviour
for a high polymer by considering the third term in equation (1)
which from the treatment of Giddings and Mallik[1] may be written in
the form

$$1/[(1/2\lambda d_p) + (D_m/w_i d_p^2 u)] \qquad (2)$$

where λ (close to unity) is a constant characteristic of the
packing, d_p is the particle diameter, D_m is assumed to be the diffusion coefficient of a molecule in free solution at infinite dilution, and w_i is a geometrical factor of order unity. For a polystyrene standard PS-20000 having a value of D_m around 10^{-6} cm^2 s^{-1}
(see reference 8), the second term in the denominator in relation
(2) at $u = 2.5$ mm s^{-1} and with $d_p = 8$ μm will be about 1% of the
first term in the denominator. We may conclude that for high
polymers the eddy diffusion term dominates mobile phase dispersion.

    It follows that only two dispersion terms, namely eddy diffusion
in the mobile phase and mass transfer in the stationary phase, have
to be considered in the expression for H for a monodisperse high
polymer. For a polydisperse solute, the expression for H should
contain a polydispersity term as suggested by Knox and McLennan.[9]
A procedure for including the true polydispersity $[\bar{M}_w/\bar{M}_n]_T$, where
$\bar{M}_w$ and $\bar{M}_n$ are the weight and number average molecular weights, has
been described, assuming the true molecular weight distribution of
the polystyrene standards is close to a logarithmic normal distribution function.[8,10] The expression for H for a permeating polydisperse high polymer is

$$H = 2\lambda d_p + [R(1-R) ud_p^2/30D_s] + (L \ln [\bar{M}_w/\bar{M}_n]_T/D_2^2 v_R^2) \qquad (3)$$

in which the second and third terms are the mass transfer and polydispersity terms respectively, where R is the retention ratio
defined by V_o/V_R, D_s is the diffusion coefficient of the solute in
the stationary phase, L is the column length, and D_2 is the slope
of the linear calibration curve of ln molecular weight against V_R
in the partial permeation range.

    In Figs. 1, 2 and 3, H for the permeating polystyrene standards
PS-3600, PS-9800 and PS-35000 increases as eluent flow rate rises
because of the mass transfer dispersion contribution. From equation (3), values of D_s may be determined from the slopes of the
curves in Figs. 1, 2 and 3 and the results are given in Table I.
Values of D_m for polystyrene standards included in Table I were
determined from the Wilke-Chang equation[11,12] and from the expression

Table I. Diffusion Coefficients of Polystyrene Standards for Silica Packings

Polystyrene	$D_m/$ 10^{-7} cm^2 s^{-1}	D_s/D_m		
		H4	SG30	γ-G SG30
PS-3600	28.9	-	0.14	0.11
PS-9800	15.8	0.19	0.13	0.10
PS-35000	7.4	0.11	0.08	0.07

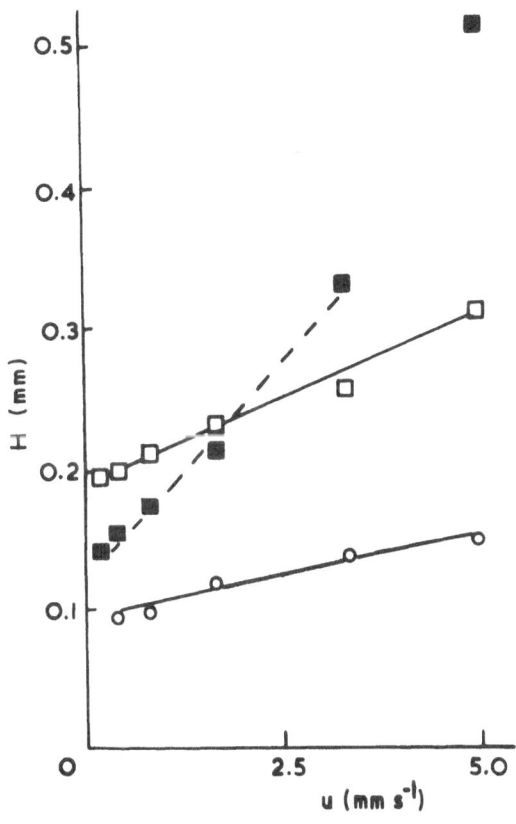

Fig. 2 Dependence of plate height on flow rate for SG30 silica; O, PS-3600; □, PS-9800; ■, PS-35000.

proposed by Rudin and Johnston.[13] The derived data for D_s in Table I are much less than values of D_m. Large errors in the procedure for determining D_s may result from the choice of value for d_p, since $d_p{}^2$ appears in the second term in equation (3). However, our silica microspheres have narrow particle size distributions. Knox and McLennan[14] have determined results for D_s by evaluating mass transfer dispersion for polystyrene standards separating with a similar porous silica ($d_p \sim 7\ \mu m$). Their results show that D_s/D_m in the range 0.06 - 0.17 increases as the molecular weight of polystyrene is reduced from 33000 to 2000. Although our calculation procedure is simpler than that employed by Knox and McLennan,[14] our results for D_s/D_m in Table I are in reasonable agreement. The observation that D_s was about $D_m/6$ reported by Giddings et al.[3] was for low molecular weight (below 5000) polystyrenes separated with much larger porous glass particles having a mean diameter of about 59 μm. Van Kreveld and van den Hoed[15] evaluated the dispersion term for mass transfer within porous silica particles (d_p = 75 - 125 μm), finding D_s/D_m to decrease from 0.31 to 0.12 as the molecular weight of polystyrene standards increased from 20000 to 160000.

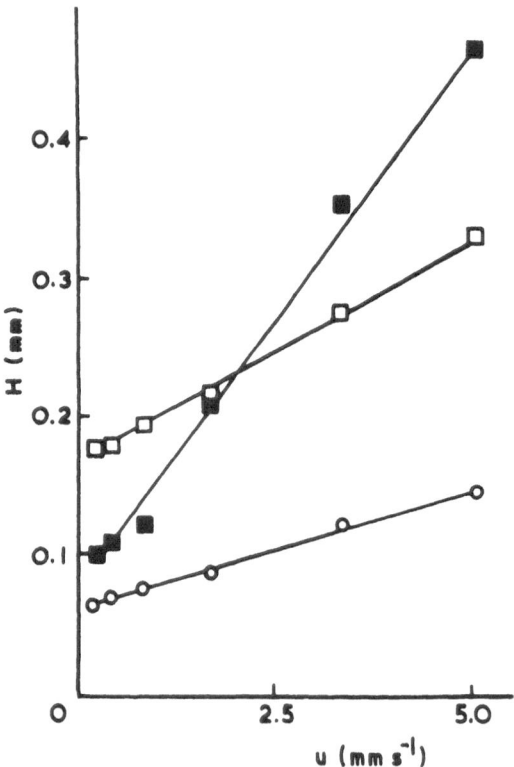

Fig. 3   Dependence of plate height on flow rate for
γ-G SG30 silica; O, PS-3600; □, PS-9800;
■, PS-35000.

CONCLUSIONS

Our results suggest that the diffusion coefficient of a high
polymer in a porous gel is much lower than the diffusion coefficient
of the same molecule in free solution at infinite dilution. Our
values of D_s/D_m are in fair agreement with data reported for polystyrene by other workers. This suggests that the simplification of
the plate height expression by assuming that the dominant mobile
phase dispersion mechanism is eddy diffusion is reasonable for a
high polymer. The low values of D_s indicate that polystyrene is
subjected to restricted diffusion in GPC for both large and small
porous particles and for packings with and without a bonded phase.

ACKNOWLEDGEMENTS

The authors wish to thank Dr.J.D.F.Ramsay and Dr.D.C.Sammon
for supplying the macroporous inorganic silicas and gratefully
acknowledge helpful discussions with Dr.R.L.Nelson, Dr.J.D.F.Ramsay
and Dr.D.C.Sammon of AERE Harwell. This research was supported by
a grant from AERE Harwell, by a CASE award from the Science Research
Council, and by a research grant from the Science Research Council
in collaboration with the SRC supported Polymer Supply and
Characterization Centre at RAPRA, Shawbury, England. The authors
thank Mr.L.J.Maisey for advice on column packing performed at the
Polymer Supply and Characterization Centre.

REFERENCES

1.   J.C.Giddings and K.L.Mallik, Anal.Chem., 38, 997 (1966).
2.   S.B.Horowitz and I.R.Fenichel, J.Phys.Chem., 68, 3378 (1964).
3.   J.C.Giddings, L.M.Bowman and M.N.Myers,
          Macromolecules, 10, 443 (1977).
4.   J.V.Dawkins and G.Yeadon, Polym.Preprints, 10 (2), 227 (1977).
5.   F.E.Regnier and R.Noel, J.Chromatogr.Sci., 14, 316 (1976).
6.   R.E.Majors, Anal.Chem., 44, 1722 (1972).
7.   P.A.Bristow, P.N.Brittain, C.M.Riley and B.F.Williamson,
          J.Chromatogr., 131, 57 (1977).
8.   J.V.Dawkins and G.Yeadon, Polymer, 20, 981 (1979).
9.   J.H.Knox and F.McLennan, Chromatographia, 10, 75 (1977).
10.   J.V.Dawkins and G.Yeadon, J.Chromatogr., 188, 333 (1980).
11.   C.R.Wilke and P.Chang, Am.Inst.Chem.Eng.J., 1, 264 (1955).
12.   R.C.Reid and T.K.Sherwood, "The Properties of Gases and
          Liquids", McGraw-Hill, New York, 1965, Chapter 8.
13.   A.Rudin and H.K.Johnston, J.Polym.Sci. Part B, Polym.Letters,
          9, 55 (1971).
14.   J.H.Knox and F.McLennan, J.Chromatogr., 185, 289 (1979).
15.   M.E.van Kreveld and N.van den Hoed, J.Chromatogr., 149, 71
          (1978).

MECHANISM OF GAS PERMEATION THROUGH POROUS POLYMERIC MEMBRANE

Kenji Kamide*, Sei-ichi Manabe*, Takashi Nohmi**,
Hiroyuki Makino**, Hiroki Narita** and Toru Kawai**

*Textile Research Lab., Asahi Chemical Industry Comp. Ltd.
Takatsuki, Osaka, Japan
**Dept. of Polymer Technology, Tokyo Institute of Technology
Ookayama, Tokyo, Japan

1 INTRODUCTION

Polymeric membrane is undoubtedly of potential industrial significance as an ultrafine separation media for food packaging, desalination of sea water, waste water treatment, artificial kidney and membrane-type artificial lung. Hitherto, the membrane utilized for the separation by the difference in the gas permeation rate have been severly limited, without exception, to those with the average pore size (diameter) less than 10 nm(=100 A).

Most investigations on the gas permeation are confined to studies of such membrane with average pore diameter, and very little work has been carried out for the membrane with sufficiently large pore, which can be recognized with an aid of electron microscope (the average diameter of the pore, defined in eq.(18), $2\bar{r}_1 < 10$nm). We define conventionally the polymeric membrane with pore of $2\bar{r}_1 \geq$ 10 nm as the porous polymeric membrane. The permeation of gas through the polymeric membrane with pore of $2\bar{r}_1 < 10$ nm has been explained in terms of dissolution-diffusion mechanism (simply termed as D flow)[1].

Up to now, the permeation of gas through the three dimensional porous media, including sintered metal and activated carbon, whose pore size is relatively large, has been studied, by simplifying the complicated gas flow in the pores with the gas flow in a bundle of uniform capillary with the same diameter, and following five kinds of flow mechanisms have been proposed so far: (1)dissolution/diffusion flow (D flow, its permeability coefficient P_D)[1], (2) viscous flow (H flow, P_H)[2], (3) slip flow (S_p flow, P_{sp})[3], (4) free molecular flow (F flow, P_F)[4], and (5) surface diffusion flow (S flow, $\mathbb{P}_S$)[5]. It should be noted that the above flow mechanisms are proposed independently for three dimensional media. No rigorous and consistent theory of flow through a porous media has been established, due to the lack of the accurate information about the complicated and irregular shape, size and the structures of the pores. With a help of the knowledge on these media, Yasuda and Tsai[6] and Cabasso et al.[7] have supposed that both H and F flows coexist in pararell in a given pore of the porous polymeric membranes. Appearance of the straight-through porous polymeric membrane in the market motivated a research for the permeation phenomena at the molecular level. This paper intends to establish a generalized theory of the permeation of gas through the porous polymeric membrane. For this purpose, at first we propose a new theory, in which a new concept of VF flow is introduced in addition to mechanisms proposed for the three dimensional media, by comparing with the literature data on the gas permeation through a single capillary and then we discuss the gas permeation through the porous polymeric membrane, on the basis of a new theory which is generalized by taking into account of the pore size distribution function $N(r)$. In addition to this, we examine the role of the chemical structure of the permeating gas on the permeation coefficient, indicating that for the membranes with small pore size ($2\bar{r}_1 < 35$ nm) S flow can never be ignored.

2 THEORETICAL BACKGROUND

2.1 Permeability Coefficients of Various Flows

The gas permeation coefficient $\mathbb{P}(P_1,P_2)$ is defined by eq.(1):

$$\mathbb{P}(P_1,P_2) = J(t)d/(P_1(t) - P_2(t)) \qquad (1)$$

where $J(t)$ is the gas flux (cm^3(STP)/$cm^2 \cdot$sec). $\mathbb{P}(P_1,P_2)$ has expressed in cm^3(STP)/cm·sec·cm Hg which is referred to as PU. The prime is attached to P when mol/$cm^2 \cdot$sec·cmHg is employed. The relation $\mathbb{P}(P_1,P_2) = \mathbb{P}(P_1,P_2)'RT_S/P_S$ holds between both permeability coefficients. $P_1(t)$ and $P_2(t)$ are the pressure (cmHg) applied at both sides ($P_1(t) > P_2(t)$) at time t.

When the above each flow exists independently in a single pore, the corresponding permeabilities are expressed by the following equations:

$$\mathbb{P}_H' = \left\{ 2\pi r^4 \bar{P}/(8\eta) \right\}/(RT) \qquad (2)$$

$$\mathbb{P}_{Sp}' = \left\{ (2-f_1)/f_1 \right\}(\pi r^3/4)(2\pi RT/M)^{1/2}/(RT) \qquad (3)$$

$$\mathbb{P}_F' = \left\{ (2-f_o)/f_o \right\}(4r^3/3)(2\pi RT/M)^{1/2}/(RT) \qquad (4)$$

$$\mathbb{P}_S' = r(\pi RT)^{3/2}(\ln P_1/P_2)/\left\{ z(P_1 - P_2)\cdot 2D_oN_A(2M)^{1/2}\cdot RT \right\} \qquad (5)$$

$$\mathbb{P}_D' = (1 - \pi r^2)DS \qquad (6)$$

where T is the measuring temperature (K), R is the gas constant, $\bar{P}$ is the average pressure $(=(P_1+P_2)/2 \ (cmHg))$, η is viscosity of the gas (poise), M is the molecular weight of the gas, f_1 is the Maxwell's reflection coefficient due to the slip flow, f_0 is the Maxwell's reflection coefficient due to the free molecular flow, $f_0=1$ is the case when the kinetic energy of the gas, colliding with the pore wall, is the same as those in equilibrium with the pore wall. In the case of $f_1=1$, the momentum of the gas molecule is transfered to the pore wall by the collision. D is the diffusion coefficient (cm^2/sec) of the gas in the polymeric materials constituting the pore membrane. S is the solubility coefficient $(mol/cm^3 \cdot cmHg)$ of the gas. z is a constant $(1.01325x10^6/76.0)$ necessary to reduce the pressure unit cmHg to CGS system of units (dyn/cm^2). D_0 is the diameter of molecule (cm). N_A is Avogadro number. It should be noted here that the place, through which the gas permeates by the solution/diffusion mechanism, is not the vacant pore itself, but the non-pore part, and P_D' in eq.(6) corresponds to the gas permeability coefficient of the membrane having a pore per unit membrane surface and accordingly, differing in the physical meaning with other parameters such as P_V', P_F' and P_S'.

When H flow occurs with S_p flow cocurrently in a given hole, we define the total flow as V flow and express its permeability with P_V. Note that S_p flow is always accompanied with H flow.

$$P_V' = P_H' + P_{Sp}' \qquad (7)$$

The special case of $f_0=f_1=1.0$ in eqs.(2) and (3) were first applied by Adzumi[2] to the flow of gas through a capillary flow. And thereafter eqs.(2) and (3) were employed for analysing the permeating phenomena of gas in three dimensional porous media[9]. Hill[11] derived the equation of the molar permeability rate under the isothermal conditions from the Sears equation[10] of the permeability rate of the gas in a circular cylinder. Hill's equation was modified as eq.(4).

2.2 Flow of Gas in a Capillary

The permeating flow of gas molecules in a circular hole can be simplified as the gas flow in a capillary. In the latter, the permeation mechanisms are determined by the relative magnitude of the mean free path of the gas molecule λ (see eq.(9)) and the pore diameter 2r: the contribution of P_F' (eq.(4)) to the total gas permeability P is predominant if $\lambda \gg 2r$ and P_H' and P_{Sp}' in eqs.(2) and (3) play an important role in P if $\lambda \ll 2r$. Both the surface diffusion and diffusion flows occur regardless of the viscous flow and the free molecular flow. If we assume that to the overall gas flow the above-described each flow contributes independently, $P(P_1,P_2)'$ can be expressed in the simple form of linear function, which does not contain any cross term,

$$P(P_1, P_2)' = f_H P_H' + f_{Sp} P_{Sp}' + f_F P_F' + f_S P_S' + f_D P_D' \qquad (8)$$

where f_H, f_{Sp}, f_F, f_S and f_D are the coefficients representing the contribution of the various flows. Obviously, $f_H = f_{Sp} = 0$ at $\lambda \geqslant 2r$ and $f_F = 0$ at $\lambda < 2r$. Eq(8) was first proposed by Nohomi et al.[13] as an extension of the equation of Adsumi and of Carman (eqs.(10) and (12)).

For the gas molecule with the molecular weight M at pressure P and at temperature T, the mean free path λ is given according to the kinetic theory of gas by

$$\lambda = \eta \, (\pi RT)^{1/2} / (2M)^{1/2} \cdot zP \qquad (9)$$

Now, we denote λ at inlet and outlet of a pore with λ_1 and λ_2, which are defined by $\lambda_1 = (\pi RT/2M)^{1/2} (\eta/zP_1)$ and $\lambda_2 = (\pi RT/2M)^{1/2} (\eta/zP_2)$, P_1 and P_2 are the pressure at inlet and outlet of a pore, respectively. For the cylindrical pore with a pore size in the range $\lambda_1 < 2r < \lambda_2$, the gas is considered to flow through a hole by the viscous flow and/or the free molecular flow mechanisms.

2.2.1 Case when the viscous (H) flow and the free molecular (F) flow occur in parallel combination

Adzumi[13] derived an expression of $P(P_1, P_2)'$ given by

$$P(P_1, P_2)' = P_H' + \gamma' P_F' \cdot f_o / (2 - f_o) \qquad (10)$$

Comparison of eq.(10) with eq.(8) shows that $f_H = 1$, $P_{Sp}' = 0$, $f_F = \gamma f_o / (2 - f_o)$, $P_S' = 0$ and $P_D' = 0$. γ' in eq.(10) is an empirical parameter and $\gamma' = 1$ at the average pressure $\bar{p} (= (P_1 + P_2)/2) = 0$. We define P_o with the pressure at which λ is equivalent with $2r$ as given by eq. (11).

$$P_o = (\eta/2zr)(\pi RT/2M)^{1/2} \qquad (11)$$

If the average pressure $\bar{P}$ is larger than P_o, γ differs depending on the gas nature (for example, 0.91 for hydrogen). The definition of H and F flows indicates that eq.(10) is theoretically inconsistent in the following points: As the gas density becomes small the probability of intermolecular collision becomes ignorable as compared with the probability of collision between the gas molecule and the pore wall, and then it can be expected H flow to be practically neglected and F flow to be a major contributor. Despite this, eq.(10) assumes that even in the above case H flow does never diminish (contradiction A). Eq.(2) implies that P_H' approaches zero as the average pressure $\bar{p}$ decreases to zero and as a result the overall permeability $P(P_1, P_2)'$ is not affected by the constancy of $f_H (= 1)$ at the limit of $\bar{P} = 0$. In the case $\bar{P} > P_o$ the probability of collision between the gas molecule and the pore wall is expected to become negligibly small, resulting in disappearance of F flow. Eq.(10) indicates that F flow does not disappear because of $\gamma > 0$ (contradiction B). The detailed experiments by Adzumi (see

Fig.8) supports strongly the theoretical prediction, denying the validity of eq.(10). In other words, in a single pore H flow and F flow do not appear in the parallel combination.

Yasuda-Tsai and Cabasso et al. have analyzed their experimental data of the gas permeation through polymeric plane membrane[6] and hollow fiber membrane[7] based on eqs.(10) and (4), where γ =1 and f_o=1 are assumed respectively. These analysis are unfortunately erroneous as will be demonstrated later.

2.2.2 Case when the viscous (H), free molecular (F) and slip (Sp) flows occur coccurrently

Carman[9] has proposed eq.(12) in place of eq.(10) by taking into account the slip (Sp) flow.

$$P(P_1,P_2)' = P_H' + (1 - \beta)P_F' + P_{Sp}' \tag{12}$$

where β is an empirical parameter and β=0 for $2r/\lambda \ll 1$ and β=1 for $2r/\lambda \gg 1$. Comparison of eq.(12) with eq.(8) shows that f_H=1, f_F =1-β, f_{Sp}=β, P_S'=0 and P_D'=0. Contradiction B in eq.(10) is overcome in eq.(12) and contradiction A in eq.(10) is not even in eq. (12), and accordingly unrealistic. Eq.(12) coincides with eq.(10) if γ' is represented by eq.(13).

$$\gamma' = \left\{(1 - \beta)(2 - f_o)/f_o + \beta(2 - f_1)/f_1\right\} \tag{13}$$

2.2.3. Case when the pressure gradient is taken into account in the gas flow through a capillary

First we assume that for $2r \leq \lambda$ F flow and for $2r > \lambda$ V flow predominate, respectively and that in the range $P_1 > P_o \geq P_2$ V flow occurs at the inlet and F flow at the outlet (this is hereafter referred to as VF flow). In other words, in a given hole V flow is combined with F flow in series combination. The overall permeability coefficient $P(P_1,P_2)'$ is given by eq.(14).

$$P(P_1,P_2)' = P_V'U(P_1 - P_o) + P_F'(1 - U(P_2 - P_o))$$
$$+ P_{VF}'U(P_2 - P_o)(1 - U(P_2 - P_o)) + f_S P_S' + f_D P_D' \tag{14}$$

where U(X) is unit step function of X and U(X)=0 at X< 0 and U(X)= 1 at X$\geq$0. In deriving eq.(8), VF flow was ignored. Comparison of eqs.(8) and (14) indicates that f_H=f_{Sp}=U(P_1 - P_o)$ and f_F=(1 - U(P_2 - P_o)).

P_{VF}' is given very approximately in the form of series combination of P_V' and P_F':

$$P_{VF}' = P_V' \cdot P_F'/\left\{(P_F' - P_V')x_o/d + P_V'\right\} \tag{15}$$

where x_o is the point of transition of V flow to F flow, as represented by the distance measured from the inlet surface, defined by

$$x_o/d = P_V'(P_1 - P_o)/[P_F'(P_o - P_2) + P_V'(P_1 - P_o)] \tag{16}$$

Under the conditions of $P(P_1,P_2)' \gg f_S P_S' + f_D P_D'$, $P(P_1,P_2)'$ can be estimated from eqs.(2), (3), (4), (14), and (15) using M, η, P_1, P_2, T, r, f_0 and f_1 data. In eq.(14) developed above V and F flow does not coexist, then, contradictions A and B in eq.(10) are reasonably overcome. Eq.(14) is a fundamental relations of a new working theory for the gas permeation through polymeric membrane.

Fig.2 shows schematic representation of the gas permeation mechanisms as functions of the distance x from the membrane surface and the pore radius r. Which flow among V, F and VF flows becomes predominant is primarily determined by the relative magnitude of the mean free path λ_1 and λ_2 (in other words, P_1 and P_2) and the pore radius. In our theory, gas flow transits from V flow to F flow at the point x_0 (shown as full line in the figure) in the range $\lambda_1/2 < r < \lambda_2/2$.

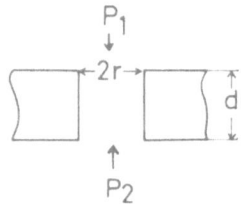

Fig.1. Cross sectional view of straight-through porous membrane; r, pore radius; P_1 and P_2, pressure loaded on both sides of the membrane

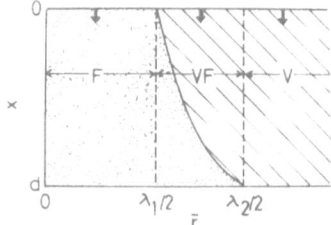

Fig.2. Gas permeation mechanisms: Arrows indicate the flow direction.

2.3 Gas Flow in a Membrane with Straight-through Cylindrical Pore

2.3.1 Permeability coefficient of the porous membrane with straight-through cylindrical pore

Fig.3 shows the schematic model of the porous membrane employed in this study. The number of pores with the pore radius between r and r + dr per unit membrane surface area is defined as $N(r)dr$, where $N(r)$ is the pore radius frequency distribution function[14]. If we can evaluate $N(r)$ for a given membrane, the pore density N ($number/cm^2$) is represented by eq.(17).

$$N = \int_0^\infty N(r)dr \qquad (17)$$

In addition, the n-th average pore radius $\bar{r}_n$ is defined as the ratio of the n-th momentum to the (n-1)-th momentum of r in the form:

$$\bar{r}_n = X_n/X_{n-1} \qquad (18)$$

with

$$X_i = \int_0^\infty r^i N(r)dr = \bar{r}_i\bar{r}_{i-1}\cdots\bar{r}_2\bar{r}_1 N \qquad (19)$$

The porosity per unit volume of the membrane P_r and the surface area of the pores existing in a unit volume S_r are given by eqs.(20) and (21), respectively.

$$P_r = \pi X_2 \qquad (20)$$

$$S_r = 2\pi X_1 \qquad (21)$$

The normal methods of measuring $N(r)$, N, $\bar{r}_n$, P_r and S_r of the porous polymeric membrane has already been fully described[14].

As is well recognized, it is unavoidable for the porous membrane to have the pore size distribution. Therefore, its overall gas

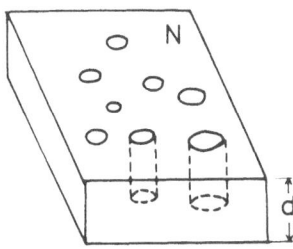

Fig.3. Straight-through porous polymeric membrane (general view): N, pore density; d, thickness.

permeability $P(P_1,P_2)'$ can be written as a straight forward generalization of the permeability coefficient for a capillary $P_c(P_1,P_2)'$

$$P(P_1,P_2)' = \int_{r_{min}}^{r_{max}} P_c(P_1,P_2)'N(r)dr \qquad (22)$$

Here $P_c(P_1,P_2)'$ is the same as $P(P_1,P_2)'$ in the preceding section. The suffix c is attached in order to distinguish a capillary flow from the gas flow through the membrane. r_{min} and r_{max} are the minimum and maximum pore radius.

Substitution of eq.(14) into eq.(22) and rearrangement of the equations, obtained thus, by using eqs.(5) and (6) leads to

$$P(P_1,P_2) = \left\{ \int_{\lambda_2/2}^{r_{max}} P_V'N(r)dr + \int_{\lambda_1/2}^{\lambda_2/2} P_V'P_F'/[(P_F'-P_V')x_o/d + \right.$$
$$P_V']N(r)dr + \int_{r_{min}}^{\lambda_1/2} P_F'N(r)dr + X_1(\ell RT)^{3/2}(\ln P_1/P_2)/$$
$$\left. [z(P_1-P_2)\cdot 2D_o\cdot N_A(2M)^{1/2}\cdot RT] + (1-Pr)DS \right\}(RT_s/P_s) \qquad (23)$$

T_s is the standard temperature (273.15 K) and P_s is the standard pressure (76 cmHg).

Eq.(23) is reduced to simplified expressions under the specific conditions:
(1) In the case when P_2 is approaching to P_1 without limit, $P(P_1,P_2)$ is given by eq.(24).

$$\lim_{P_2 \to P_1} P(P_1,P_2) = P(P_1,P_1)$$
$$= \left\{ \int_{\lambda_1/2}^{\infty} P_V'N(r)dr + \int_0^{\lambda_1/2} P_V'N(r)dr \right.$$
$$+ X_1(\ell RT)^{3/2}/zP_1\cdot 2D_oN_A(2M)^{1/2}\cdot RT$$
$$\left. + (1-Pr)DS \right\}T_s/P_s\cdot T \qquad (24)$$

At the limit of $P_2 \to P_1$, eq.(8) coincides with eq.(24).
(2) In the case when $P_1 \to 0$ (accordingly, $P_2 \to 0$) and the adsorption velocity of the gas onto the pore wall is much larger than its desorption velocity (in other words, adsorption isotherms are not attained yet), the third term of the left-hand side of eq.(24) becomes almost zero and then

$$P(0,0) = \left\{ [(2-f_o)/f_o][4X_3/3(2\ell RT/M)^{1/2}] + (1-Pr)DS \right\}T_s/P_sT \qquad (25)$$

In so far as the polymeric membranes employed in this study are concerned, $(1-Pr)DST_s/P_sT$ can be ignored if it is less than 10^{-9}(PU).

When an inorganic gas permeates through the polymeric membrane with $2\bar{r}_1 \gtrsim 0.01 \mu m$, the contribution of $\mathbb{P}_S'$ and $\mathbb{P}_D'$ to $\mathbb{P}(P_1,P_2)'$ in eq.(23) can be neglected. Then, eq.(23) can be rewritten as

$$\mathbb{P}(P_1,P_2) = [\int_{\lambda_2/2}^{r_{max}} \mathbb{P}_V' N(r) dr + \int_{\lambda_1/2}^{\lambda_2/2} \mathbb{P}_{VF}' N(r) dr$$

$$+ \int_{r_{min}}^{\lambda_1/2} \mathbb{P}_F' N(r) dr] RT_s/P_s \qquad (26)$$

The permeation mechanism in a porous polymeric membrane can be classified into the following six categories, depending on the relative magnitude of the pore size r_{max}, r_{min} and the mean free path (λ_1 and λ_2). Fig.4 demonstrates the classified patterns. (1) Case A: $2r_{min} > \lambda_2$, (2) Case B: $\lambda_1 < 2r_{min} \leqslant \lambda_2$ and $2r_{max} > \lambda_2$, (3) Case C: $2r_{min} \leqslant \lambda_1$ and $2r_{max} > \lambda_2$, (4) Case D: $\lambda_1 < 2r_{min} \leqslant 2r_{max} < \lambda_2$, (5) Case E: $2r_{min} \leqslant \lambda_1 < 2r_{max} \leqslant \lambda_2$, (6) Case F: $2r_{max} \leqslant \lambda_1$. Then, $\mathbb{P}(P_1,P_2)$ or $\mathbb{P}(P_1,P_1)$ for the above cases will be given below.

2.3.2 Case A

The second and third terms in the right-hand side of eq.(26) are zero and eq.(26) is simplified into:

$$\mathbb{P}(P_1,P_2) = [(z\pi X_4/8\eta)\bar{P} + \{(2 - f_1)/f_1\} (\pi X_3/4) (2\pi RT/M)^{1/2}]x$$

$$(T_s/P_s T) \qquad (27)$$

From eq.(27) we obtain

$$\lim_{P_2 \to P_1} \mathbb{P}(P_1,P_2) = \mathbb{P}(P_1,P_1) = [(z\pi X_4/8\eta)P_1 + \{(2 - f_1)/f_1\} (\pi X_3/4)x$$

$$(2\pi RT/M)^{1/2}](T_s/P_s T) \qquad (28)$$

In case A, the experimental results show that the pressure dependence of f_1 in eq.(3) can be ignored (see Fig.24). Therefore, the intercept of the plot of $\mathbb{P}(P_1,P_1)$ against P_1 gives $\{(2-f_1)/f_1\}$. $(\pi X_3/4)(2\pi RT/M)^{1/2}(T_s/P_s \cdot T)$ and its slope yields $(z\pi X_4/8\eta)T_s/P_s \cdot T$.

2.3.3 Case B

The third term in the right-hand side of eq.(26) is zero. $\mathbb{P}(P_1,P_1)$ is given by eq.(28) and $\lim_{P_2 \to 0} \mathbb{P}(P_1,P_2)$ is given by eq.(34) in case D

2.3.4 Case C

Application of the following two approximations such as

$$\int_{r_{min}}^{\lambda_1/2} r^4 N(r) dr = \lambda_1 \alpha_1 X_3/4 \qquad (29)$$

and

$$\int_{\lambda_1/2}^{r_{max}} r^4 N(r)\,dr = (\lambda_1/2 + r_{max})(1 - \alpha_1)X_3/2 \tag{30}$$

to eq.(28) leads us to eq.(31)

$$\mathbb{P}(P_1,P_1) \doteqdot [\{1 - \frac{\alpha_1}{1 + k_1(1-\alpha_1)}\}\frac{z\pi P_1}{8\eta}\,X_4$$

$$+ (\frac{2\pi RT}{M})^{1/2}\{\frac{2 - f_1}{f_1}\frac{\pi}{4}(1 - \alpha_1)$$

$$+ \frac{2 - f_o}{f_o}\frac{4}{3}\alpha_1\}X_3]\,T_s/P_s \cdot T \tag{31}$$

where
$$\alpha_1 = \int_{r_{min}}^{\lambda_1/2} r^3 N(r)\,dr / \int_{r_{min}}^{r_{max}} r^3 N(r)\,dr \tag{32}$$

$$k_1 = 2r_{max}/\lambda_1 \tag{33}$$

α_1 exhibits λ_1 dependence (accordingly, P_1 dependence) and in this pressure range the P_1 dependence of neither f_o nor f_1 can also be neglected. For this reason, the plot of $\mathbb{P}(P_1,P_1)$ vs. P_1 is not always represented by a straight line. $\mathbb{P}(P_1,0)$ is given by eq.(35) in case E.

2.3.5 Case D

Both the first and third terms in the right-hand side of eq. (26) become zero and $\mathbb{P}(P_1,P_2)$ is expressed by eq.(34)

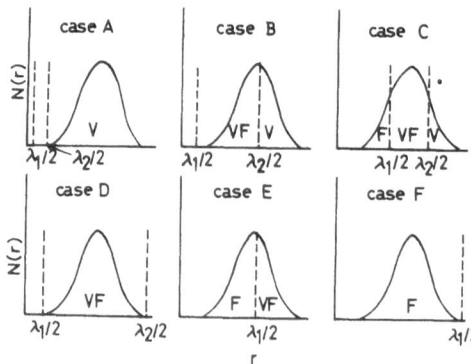

Fig. 4. Classified patterns of gas permeation mechanisms for porous membrane having distribution of pore size.

$$P(P_1,P_2) = \int_{r_{min}}^{r_{max}} P_{VF}'N(r)dr \cdot (RT_s/P_s) \qquad (34)$$

2.3.6 Case E

The expression for $P(P_1,P_2)$ is obtained by putting the first term in the right-hand side of eq.(26). $P(P_1,P_1)$ is given by eq. (31) in case C. $P(P_1,0)$ is represented in the form:

$$P(P_1,0) + [\int_{\lambda_1/2}^{r_{max}} P_{VF}'N(r)dr \cdot RT$$

$$+ (\frac{2 - f_o}{f_o})(\frac{2\pi RT}{M})^{1/2}(\frac{4}{3})\alpha_1 x_3]T_s/P_s T \qquad (35)$$

2.3.7 Case F

Both the first and second terms in the right-hand side of eq. (26) become zero and $P(P_1,P_2)$ is given by

$$P(P_1,P_2) = [(2 - f_o)/f_o](4/3)(2\pi RT/M)^{1/2}x_3(T_s/P_s T) \quad (36)$$

In this case, if f_o is independent of the pressure, eq.(36) reduces to

$$P(P_1,P_2) = P(P_1,P_1) = P(P_1,0) = P(0,0) \qquad (37)$$

Case F is suitable for investigating the detail of F flow. Eqs.(27) − (37) reveal that we can evaluate separately V flow from $P(P_1,P_2)$ in case A, F flow from $P(P_1,P_1)$ in case F and VF flow from

$P(P_1,P_2)$ in case D.

3. EXPERIMENTAL

3.1 Membrane

Commercially available polycarbonate membranes "Nuclepore" (General Electric Corp.) Nu0.8, Nu0.6, Nu0.2, Nu0.1, Nu0.08, Nu0.05, Nu0.03, Nu0.015 and non-alkalitreated membrane Nu0.00 were utilized. In order to remove parafines attached to Nuclepore membrane surface, the membrane as received was washed at 20 °C with diethyl ether, dried in vacuo. Nuclepore has straight-through pores and highly suitable for the methematical analysis. Homogeneous polycarbonate membrane Nu_s was cast from chloroform solution. Of course, Nu0.00 Nu_s are outside of a categories of the porous membrane, but they are utilized to investigate the role of P_S and P_D.

The porosity Pr was determined by the apparent density method. N(r) was evaluated by the scanning electron microscopic method except

for Nu0.8, for which the mercury intrusion method was applied[15].
The experimentally obtained N(r) data in lower r region becomes
less reliable. The absolute value of r, determined by the electron
microscopic method, was corrected so as to equalize the X_4 value,
calculated from N(r) by the electron microscopic method, with the
corresponding value determined by the water filtration rate method
[15]. The correction factor was always less than 10 %. The maximum
pore radius r_{max} was determined by the bubble point method[15].
Table 1 summerizes N, $\bar{r}_i$ (i=1 - 4), r_{max}, r_{min}, Pr, Sr, X_i (i=1 -
4) for Nuclepore membranes. Fig.5 illustrates N(r) for porous polycarbonate membranes.

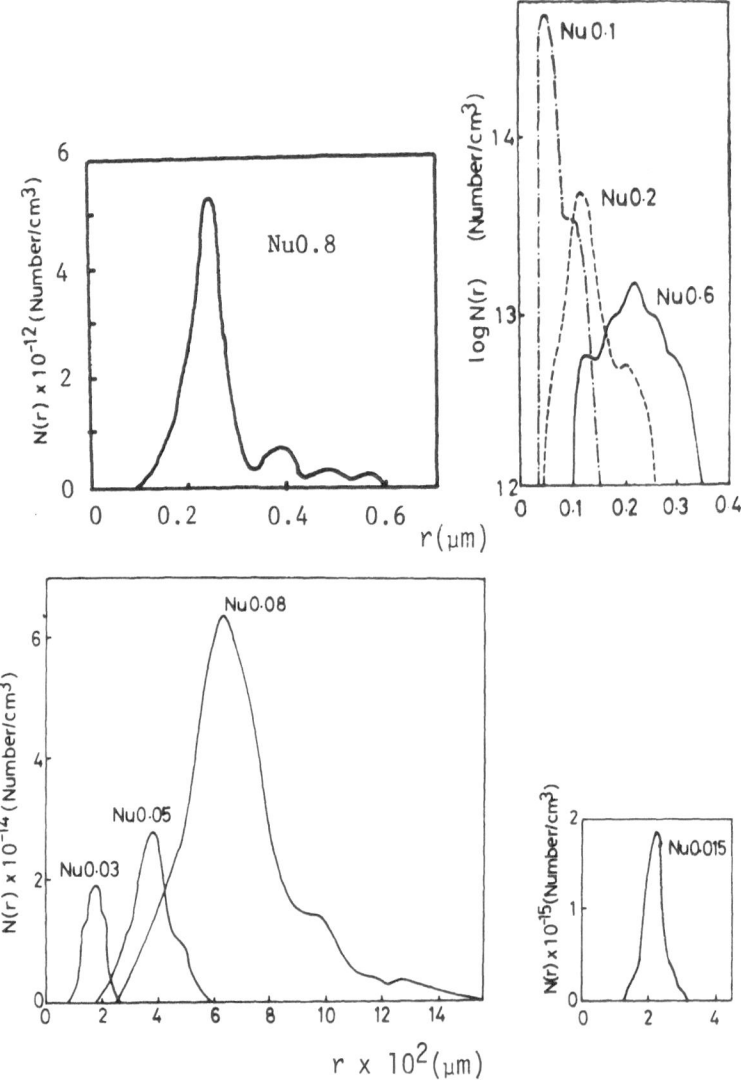

Fig.5 Pore size distribution function N(r) of polycarbonate membrane
      "Nuclepore" deternined by electron scanning microscope.

Table 1. Characterization of pore of polycarbonate membrane "Nuclepore".

Characteristic value		Nu0.8	Nu0.6	Nu0.2	Nu0.1
N (number/cm^2)		5.42×10^7	9.1×10^7	2.1×10^8	8.0×10^8
$\bar{r}_i$	$i=1$ (μm)	0.356	0.235	0.118	0.0629
	$i=2$ (μm)	0.456	0.252	0.132	0.0715
	$i=3$ (μm)	0.501	0.268	0.156	0.0840
	$i=4$ (μm)	0.567	0.325	0.183	0.0984
r_{max} (μm)		0.580	0.34	0.19	0.11
r_{min} (μm)		0.10	0.10	0.045	0.028
Pr $(-)$		0.22	0.14	0.13	0.10
X_i	$i=1$ (cm^{-1})	1.93×10^3	2.15×10^3	2.48×10^3	5.05×10^3
	$i=2$ ($-$)	6.87×10^{-2}	5.41×10^{-2}	3.26×10^{-2}	3.61×10^{-2}
	$i=3$ (cm)	3.45×10^{-6}	1.45×10^{-6}	5.09×10^{-7}	3.03×10^{-7}
	$i=4$ (cm^2)	1.95×10^{-10}	4.71×10^{-11}	9.31×10^{-12}	$2.99 \times 10{-12}$

Characteristic value		Nu0.08	Nu0.05	Nu0.03	Nu0.015
N (number/cm^2)		6.83×10^8	3.91×10^8	1.62×10^7	1.03×10^9
$\bar{r}_i$	$i=1$ (μm)	5.60×10^{-2}	4.59×10^{-2}	1.74×10^{-2}	2.21×10^{-2}
	$i=2$ (μm)	5.99	4.71	1.80	2.21
	$i=3$ (μm)	6.56	4.75	1.86	2.27
	$i=4$ (μm)	7.13	4.96	1.92	2.32
r_{max} (μm)		18.2×10^{-2}	6.6×10^{-2}	3.4×10^{-2}	3.2×10^{-2}
r_{min} (μm)		2.5	1.8	0.7	1.3
Pr $(-)$		0.072	0.026	0.020	$-$
X_i	$i=1$ (cm^{-1})	3.83×10^3	1.80×10^3	2.82×10	2.35×10^3
	$i=2$ ($-$)	2.29×10^{-2}	8.45×10^{-3}	5.09×10^{-5}	5.20×10^{-3}
	$i=3$ (cm)	1.50×10^{-7}	4.01×10^{-8}	9.46×10^{-11}	1.18×10^{-8}
	$i=4$ (cm^2)	1.07×10^{-12}	1.99×10^{-13}	1.81×10^{-16}	2.74×10^{-14}

3.2 Permeating Gas

Helium and argon as single atom molecule gas, hydrogen, nitrogen, oxygen and carbon monooxide as two atom molecule gas, carbon dioxide as three atom molecule gas, organic gas such as C_2H_2, C_2H_4, C_3H_6, C_4H_6, C_3H_8 and iso-C_4H_{10}, manufactured by Japan oxygen Co., were employed. The purity of these gases was above 99.9%. To remove impurity component, the organic gas was liquidized by cooling with liquid nitrogen before use.

3.3 Measuring Instrument of Gas Permeability Coefficient

The permeability coefficient was determined for $P_2 \gtrless 76$ cmHg by using an apparatus shown in Fig.6 and for $P_2 < 76$ cmHg by using an apparatus shown in Fig.7, both for specially designed use and constructed in our laboratory.

On PF2 in the instrument in Fig.7, the porous membrane with the same pore size as that, whose permeability is to be measured, was inserted in order to remove the fine dust floating in the permeating gas. The supporting material for the filter holder was designed to have larger hole so as to allow us to neglect the effect of the hole of holder. Under the conditions of $P_1-P_2 < 380$ cmHg, the pore size of the porous membrane was not significantly influenced by the applied pressure P_1. These instruments were installed in a bath controlled to within $\pm$ 0.5 °C and the measurements were made at 25 °C if not described.

3.4 Permeation Experiment

Experimental procedure of the gas permeation is described below in the case of the instrument in Fig.7.

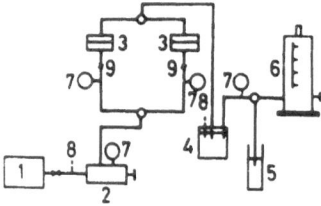

Fig.6. Apparatus for measurement of the J-ΔP curve by the combined bubble pressure and fluid permeability method: 1, pressure source (compressor); 2, pressure control handle; 3, filter holder; 4, trap for solvent and oil; 5, bubble tester; 6, flow meter; 7, pressure gauge; 8, thermometer; 9, safety valve.

Measuring method 1:   The specimen gas flowed from the bomb through
a drying tube packed with calcium chloride and phosphorous pentoxide
into the storage tank A, which was in advance evacuated.   The pressure at the storage tank $P_1(t)$ was adjusted to be constant (5 - 70
cmHg).   After the temperature of the gas in the storage tank became
constant, the permeation experiment started.   The gas flow flowed
from the tank A, through the porous membrane into the storage tank
B.   The pressure of the gas in the tank B $P_1(t)$ was measured as a
function of time t by mercury manometer M2.   The permeability coefficient $P(P_1, P_2)$ was calculated by using eq.(44) from $P_2(t)$ data.

Measuring method 2:   After introduction of the gas in the tank A,
the gas permeated through the porous membrane in the same manner as
method 1.   During the gas permeation, the tank B was continuously
evacuated by vacuum pump in order to maintain $P_2(t)$ constant.   The
variation of $P_1(t)$ was measured by mercury manometer M1.   For the
permeation experiment of mixed gases, method 1 was employed.   In
this case, a Hitachi gas chromatography (model K23) was directly
connected to the instrument in Fig.7, and the composition of the
permeated gas was analyzed.

4   ANALYSIS OF EXPERIMENTAL DATA

4.1 Method for Evaluating Gas Permeation Coefficient

      In the case of method 1, the experimental $P_2(t)$ data are expan-
ded in power series of time t up to 6th order:

$$P_2(t) = a_0 + a_1 t + a_2 t^2 + \cdots + a_6 t^6 \tag{39}$$

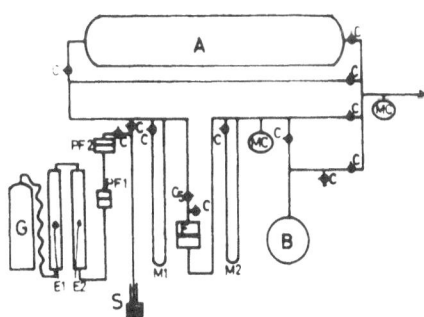

Fig.7. Measuring instrument of gas permeability coefficient:
       A, storage tank of gas of high pressure side; B, storage
       tank of gas of low pressure side; C and C5, switch cock; E1
       and E2, drying tubes containing $CaCl_2$ and P_2O_5, respectively;
       F, filter holder; G, gas bomb; M1 and M2, mercury manometer;
       PF1, prefilter with ceramic filter; PF2, prefilter with Nu0.8
       MC, Macleod gauge.

The coefficients a_0, $\cdots$, a_6 are estimated by the least-square method. $P_1(t)$ is calculated by eq.(40)

$$P_1(t) = P_1(0) - [V_2(t)/V_1(t)]P_2(t) \qquad (40)$$

with

$$V_1(t) = B_1 P_1(t)/2.0 + V_1(0) \qquad (41)$$

and

$$V_2(t) = B_2 P_2(t)/2.0 + V_2(0) \qquad (42)$$

$V_1(t)$ and $V_2(t)$ are the volume of the storage tanks A and B, B_1 and B_2 are the cross section of mercury manometers M1 and M2, respectively.

The gas flux $J(t)$ is calculated by putting $P_2(t)$ in eq.(39) and $V_2(t)$ in eq.(42) into eq.(43)

$$J(t) = [d(V_2(t)P_2(t)/RT)/dt](1/A)(RT_s/P_s)$$

$$= (V_2(t)T_s/AP_s T)(dP_2(t)/dt) \qquad (43)$$

Combination of eq.(1) and eq.(43) leads to

$$\mathbb{P}(P_1,P_2) = \frac{T_s}{P_s T}V_2(t)\frac{d}{A}\cdot\frac{1}{(P_1(t)-P_2(t))}\frac{dP_2(t)}{dt} \qquad (44)$$

where d is the thickness of the membrane, A is the effective permeation area. In method 1, $P_1(0) = P_1(t)$ holds throughout the measurements.

In method 2, $P_1(t)$ data are expressed by eq.(45).

$$\mathbb{P}_1(t) = b_0 + b_1 t + b_2 t^2 + \cdots + b_6 t^6 \qquad (45)$$

The coefficients b_0, $\cdots$, b_6 are determined by the least-square methods. $\mathbb{P}(P_1,P_2)$ is evaluated by eq.(46) from $P_1(t)$ in eq.(45) and $P_2(t)$ ($P_2(0)$).

$$\mathbb{P}(P_1,P_2) = \frac{T_s}{P_s T}V_1(t)\frac{d}{A}\cdot\frac{1}{P_1(t)-P_2(0)}\cdot\frac{dP_1(t)}{dt} \qquad (46)$$

4.2 Determination of Apparent Activation Energy E_p of Gas Permeation

Apparent activation energy E_p is obtained by the temperature dependence of $\mathbb{P}(P_1,P_1)$:

$$E_p = -R\cdot d\ln \lim_{P_2 \to P_1} \mathbb{P}(P_1,P_2)/d(1/T) = -R\cdot d\ln \mathbb{P}(P_1,P_1)/d(1/T) \qquad (47)$$

Up to now, $P(\bar{P})$ was utilized in place of $P(P_1,P_1)$ in eq.(47).

5 RESULTS AND DISCUSSION

5.1 Permeation of Gas Through Capillary Having Constant Pore Size

As early as 1937, Adzumi has carried out an outstanding pion-eering work on the gas flow in a glass capillary. He has calculated $P(P_1,P_2)$ by assuming that it is dependent of $\bar{P}$ alone. This assump-tion is not acceptable. The experimental Adzumi's data on $P(0,0)$ for H_2, C_2H_2 and C_3H_6 at 20 °C are analyzed according to the theory presented. As a result, following relations are obtained.

$$P(0,0) = P_F \quad (eq.(4) \text{ with } f_o=1)$$

$$dP(P_1,P_2)/d\bar{P} = \pi r^4/4\eta RT$$

These findings suggest that F flow alone occurs at $2\bar{r} \ll \lambda$ (r; capi-llary radius). Adzumi has analyzed his data according to eq.(10) and obtained $P(0,0)$ and $dP(P_1,P_2)/d\bar{P}$ values.

Next, we examine the flow coexisting with H flow at $2\bar{r} \gg \lambda$. The difference between the experimental $P(P_1,P_2)$ obtained by Adzumi and the theoretical $P_H(=P_H'RT_s/P_s)$(eq.(2)), $P(\bar{P}_1,P_2)-P_H$ is almost independent of $\bar{P}$, being proportional to $M^{-1/2}$. The latter means that at $2\bar{r} \gg \lambda$ Sp flow and/or F flow exist significantly. We can easily judge which flow is predominant by analyzing the composition of binary gas mixture before and after the permeation experiment. That is, the separation coefficient α defined by $(\lambda_1/\lambda_2)_o/(\lambda_1/\lambda_2)_i$ (λ_1 and λ_2 are molar concentrations of components 1 and 2, the su-ffix o and i mean the outlet and inlet of the capillary) is expected theoretically to be $(M_2/M_1)^{1/2}$ in F flow and to be 1 in Sp flow, where M_2 and M_1 are the molecular weight of the gas components 2 and 1, respectively. In fact, by analysis of Adzumi's data (Table 3 of ref.14)on H_2/C_2H_2 mixture α at $\bar{P} \rightarrow 0$ is found to be 3.60, which agrees well with the theoretical value $(M_2/M_1)^{1/2} = (26/2)^{1/2}$ =3.61 of F flow, and α at $\bar{P} > 1$ mmHg was 1.0. Then, we can conclude that at $\bar{P} \rightarrow 0$ F flow alone occurs and in the pressure range, in which all gas molecules flow on the H flow mechanism, F flow does not co-exist, in other words, the flow which coexists with H flow and is proportional to $M^{-1/2}$ is Sp flow. Above-mentioned experimental fa-cts can not be explained by eq.(10).

Fig.8 shows the dependence of $P(P_1,P_2)$ on $\bar{P}$ (full line). This relation was calculated by using eqs.(2)-(4), (14)-(16) from $\bar{r}$, T, η, M data (in Adzumi's experiments) and assuming $f_o=1$ and $f_S=f_D=0$. Here, f_1 value was chosen in the manner that the experimental $P(P_1, P_2)$ value coincide with theoretical one at $\bar{P}=2$mmHg. For example f_1 was found to be 0.79 for H2, 0.80 for C_2H_2 and 0.79 for C_3H_6. In the figure, $P_1=10P_2$ and the independence of f_o and f_1 on the pre-ssure were assumed. Whole experimental data points by Adzumi as

shown as open circle are evidently consistent with the theoretical
curve, based on eq.(14). In this manner, eq.(14) is experimentally
as well as theoretically by far adequate than eqs.(10) and (12) for
a better understanding of $P(P_1,P_2)$ of the gas flow in a capillary.
In eq.(14) is not included the parameters less clear physical meaning such as γ and β. By use of eq.(14) the quantitative relations
of the porous polymeric membrane between $P(P_1,P_2)$ and $N(r)$ can be
obtained[15] and serve as a method for estimating $N(r)$ from $P(P_1,P_2)$
data. Adzumi attempted to explain the minimum in Fig.8 by eq.(10)
and met with failure. An appearance of the minimum observed in
Fig.8 can be explained by eq.(14) in terms of a new occurrence of V
flow in the vicinity of the inlet of a capillary because of the approaching of λ_1 to $2\bar{r}$ with an increase in P_1.

From the numerical analysis of eq.(14) we can conclude that (1)
below the pressure, at which the $P(P_1,P_2)$ vs. $\bar{P}$ plot began to devi-
ate from the straight line drawn from higher $\bar{P}$ region (for example,
$\bar{P} > 0.1$ cmHg in Fig.8), $\alpha_0 \neq 1$, (2) a minimum of $P(P_1,P_2)$ appears
around $\bar{P}$ at which $2\bar{r} \doteqdot \lambda$ is satisfied, (3) strictly speaking, the
$\lambda/2\bar{r}$ value at minimum $P(P_1,P_2)$ is larger as f_1 is larger, (4) as
far as $(4/3)(2 - f_0)/f_0 - (\pi/4)(2 - f_1)/f_1 > \pi/32$ is satisfied, a minimum $P(P_1,P_2)$ is expected to appear at $2\bar{r} \doteqdot \lambda$. (1) - (3) can also
be confirmed experimentally by detailed examination of Adzumi's
data.

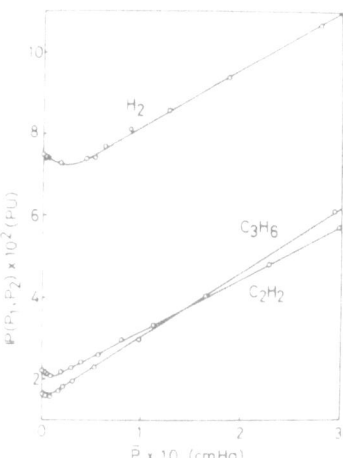

Fig.8. Mean pressure $\bar{P}$ dependence of gas permeability coefficient
$P(P_1,P_2)$ of a capillary ($2\bar{r}=2.42 \times 10^{-2}$cm, d=8.7 cm) for various gases at T=293.15K($P_1=10P_2$):
o, observed value by Adzumi[2]; full line, calculated value by
eq.(14); M=2.0, η =0.89x10^{-4} poise and f_1=0.79 for H$_2$; M=26.0
η=1.022x10^{-4}poise and f_1=0.80 for C$_2$H$_2$; M=46.0, η =0.848x10^{-4}
poise and f_1=0.79 for C$_3$H$_6$.

5.2 Permeation of Inorganic Gas Through Porous Polycarbonate Membrane

Fig.9 - Fig.18 shows $\bar{P}$ dependence of $P(P_1,P_2)$ for various poly-carbonate membranes. In these figures, broken line is $P(P_1,P_1)$ and dotted line is $P(P_1,0)$, both estimated from the experiments. Evidently, $P(P_1,P_2)$ is not a function of a single variable $\bar{P}$, but is not a function of a single variable $\bar{P}$, but is dependent of both P_1 and P_2. $P(P_1,P_1) > P(2P_1,0)$ holds even if $\bar{P}$ is kept constant (P_1+P_2) $=(2P_1+0)/2=P_1$. When an organic gas penetrates through a membrane, the relative magnitude between $P(P_1,P_1)$ and $P(2P_1,0)$ often reverses, in particular as the pore size decreases.

5.2.1 CaseA: When the gas flow through porous membrane is V flow alone

Fig.9 show the plot of $P(P_1,P_2)$ against $\bar{P}$ for Nu0.8. The data points were obtained by using the instrument shown in Fig.6 (in the case of Fig.9(a)) and the one in Fig.7 (Fig.9(b) and (c)). At P = 76 cmHg λ is 0.0653 μm for air at 25 °C (eq.(9)). Therefore, $P_2 \gtrsim$ 76 cmHg is equivalent to $2r_{max} < \lambda_1$ indicating the experimental data in Fig.9(a) is just the case A. Since P_D is less than 10^{-9}(PU) and P_S is ca. 2×10^{-6}(PU) for Nu0.8, contribution of both P_D and P_S to the overall $P(P_1,P_2)$ can be ignored (eq.(14)). Full line in Fig.9 (a) is the curve, calculated from eq.(27) by using $f_1=0.87$, which was determined in the manner that the experimental $P(76,76)$ coincides with theoretical one. The agreement of $P(P_1,76)$ between the theory and the experiment is reasonable. At $2r_{max} < \lambda_1$, that is in case A, eq.(27) can explain the experimental $P(P_1,P_2)$ and f_1 does not depend P_1.

Now, the $\bar{P}$ dependence of $P(P_1,P_2)$ can be empirically expressed by

$$P(P_1,P_2) = C_1\bar{P} + C_2 \tag{48}$$

Substitution of eq.(27) into eq.(48) yields

$$\bar{r}_4 = (C_1/C_2)[(2 - f_1)/f_1](2\eta/z)\cdot(2\pi RT/M)^{1/2} \tag{49}$$

Putting experimental C_1 and C_2 and $f_1=0.87$ (employed before) into eq.(49) we obtain $\bar{r}_4=5.56\times10^{-5}$ cm for Nu0.8. This is in excellent coincidence with the experimental value(5.67×10^{-5} cm) by scanning electron microscopy (see Table 1). It should be noted that the coefficient attached to (C_1/C_2) and the physical meaning of C_2 in eq. (49) is different from the corresponding equation derived by Yasuda and Tsai[6]. They employed

$$P(P_1,P_2) = N[z\pi r^4\bar{P}/8 + (4\pi r^3/3)\cdot(2\pi RT/M)^{1/2}](T_sN_A/P_sT) \tag{50}$$

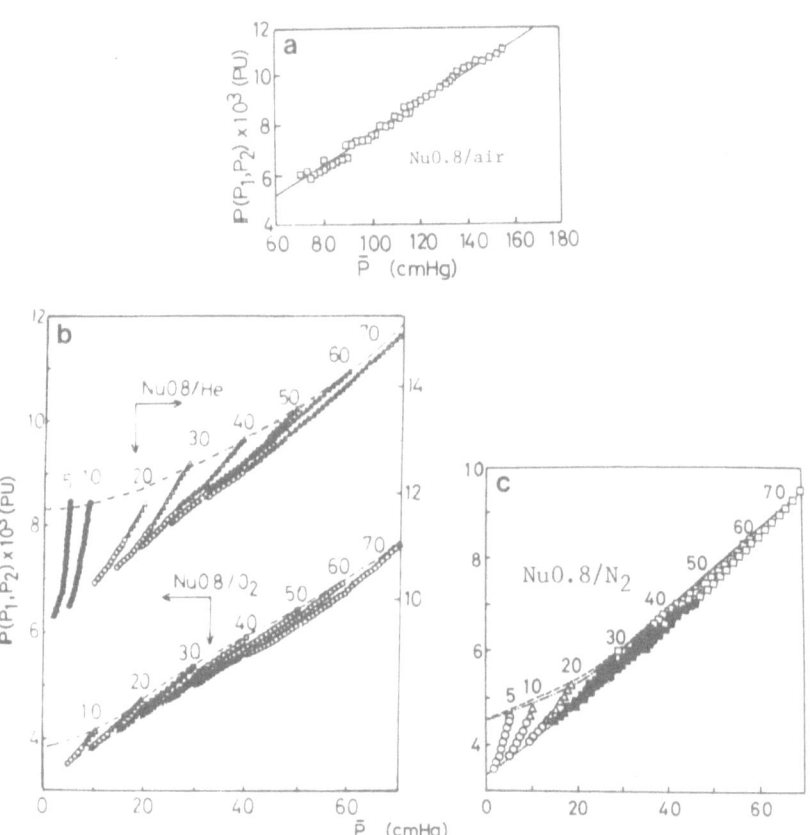

Fig.9. Mean pressure P̄ dependence of gas permeability coefficient
P(P_1,P_2) of Nu0.8 for various gases:
All marks are experimental data points, ●, case F; o, case
E; ▲, case D; △, case C; ■, case B; □, case A (see Fig.4);
——, calculated value by using eq.(27); —..—, calculated
value by using eq.(24); ----, P(P_1,P_1); ····, P(P_1,0); numbers on the curves mean P_1 (cmHg).

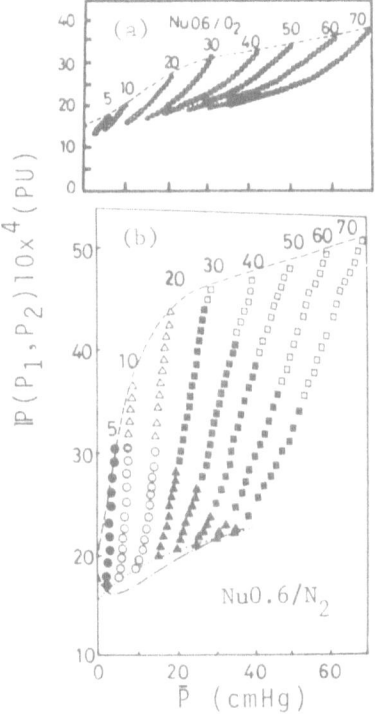

Fig.10. $\bar{P}$ dependence of $P(P_1,P_2)$ of Nu0.6 for O_2 and N_2:
         The meanings of all marks and lines are same as Fig.9;
         chain line indicates $P(P_1,0)$ obtained by method 2 in the
         text.

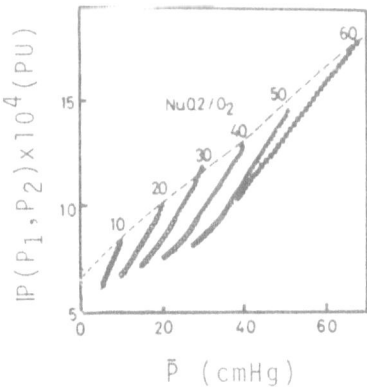

Fig.11. $\bar{P}$ dependence of $P(P_1,P_2)$ of Nu0.2 for O_2:
         The meanings of all marks and lines are same as Fig.9.

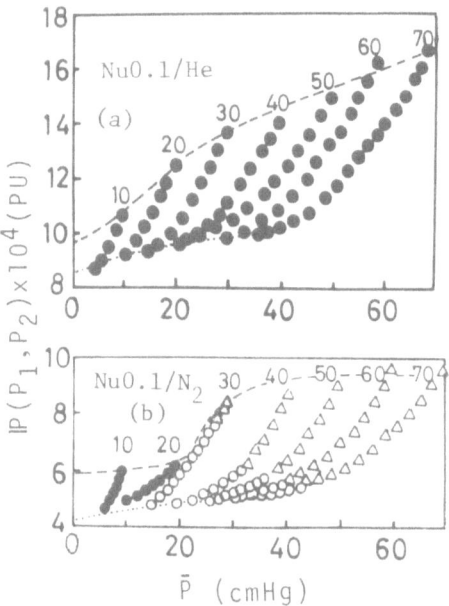

Fig.12. $\bar{P}$ dependence of $P(P_1,P_2)$ of Nu0.1 for He((a)) and N_2((b)):
The meanings of all marks and lines are same as Fig.9.

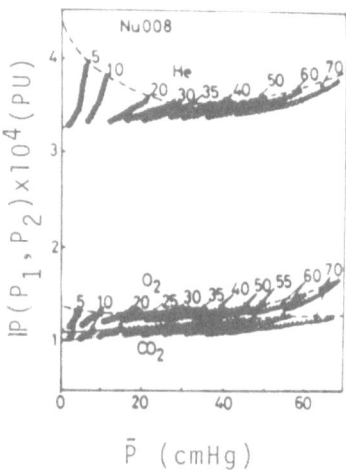

Fig. 13 $\bar{P}$ dependence of $P(P_1,P_2)$ of Nu0.08 for He, O_2, and CO_2:
The meanings of all marks and lines are same as Fig.9.

Fig.14. $\bar{P}$ dependence of $\mathbb{P}(P_1,P_2)$ of Nu0.05 for He, O_2, and CO_2:
The meanings of all marks and lines are same as Fig.9.

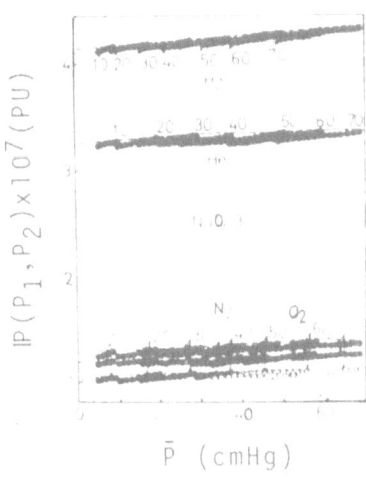

Fig.15. $\bar{P}$ dependence of $\mathbb{P}(P_1,P_2)$ of Nu0.03 for H_2, He, N_2, O_2, and CO_2:
The meanings of all marks and lines are same as Fig.9.

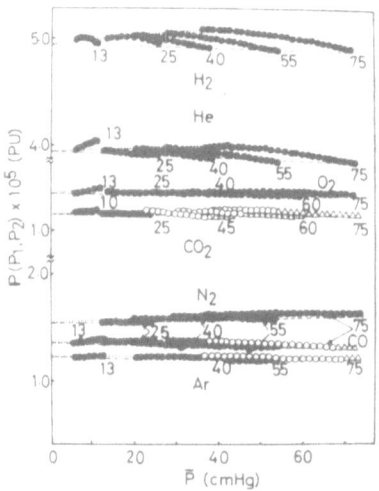

Fig.16. P̄ dependence of ℙ(P₁,P₂) of Nu0.015 for H_2, He, O_2, CO_2, N_2
        CO, and Ar:
        The meanings of all marks and lines are same as Fig.9.

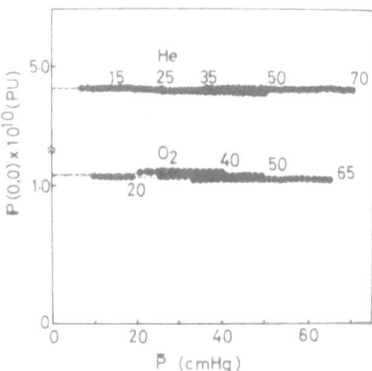

Fig.17. P̄ dependence of ℙ(P₁,P₂) of Nu0.00 for He and O_2:
         The meanings of all marks and lines are same as Fig.9.

In deriving eq.(50) the coexistence of V flow and F flow is assumed and this assumption is, as noted before, unacceptable.

Parameters C_1 and C_2 were evaluated from the experimental $P(P_1, P_2)$ for various membranes, belonging to case A. Figs. 19 and 20 show the plot of C_1 vs. $X_4/8\eta$ and that of C_2 vs. $X_3/M^{1/2}$. The full lines are theoretical curves calculated by eqs.(51) and (52)

$$C_1 = z \pi (X_4/8\eta)(T_s/P_s T) \tag{51}$$

and

$$C_2 = [(2 - f_1)/f_1](X_3/M)^{1/2}(\pi/4)(2\pi RT)^{1/2}(T_s/P_s T) \tag{52}$$

with

$$f_1 = 0.87 - a/(P_1 + b) \tag{53}$$

$a=5$ cmHg, $b=18$ cmHg.

Figs.19 and 20 show the validity of eqs.(52) and (53), in other words, in case A the permeability coefficient is given by eq.(27).

5.2.2 Case F: When the gas flow through porous membrane is F flow alone

$P(P_1,P_2)$ data, corresponding to case F in Figs.9 - 18, designated by closed circle show the very slight $\bar{P}$ dependence. In case F, eq.(36) indicates that if f_o is absolutely pressure independent, $P(P_1,P_2)$ does not depend on $\bar{P}$. The slight $\bar{P}$ dependence experimentally observed is therefore due to the pressure dependence of f_o. The value of $\lim_{P_1 \to 0} P(P_1,P_1) = P(0,0)$ was extrapolated from the case F data, and was plotted as a function of $X_3 M^{-1/2}$ in Fig.21. Full line

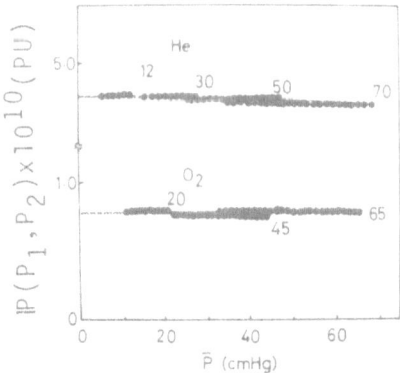

Fig.18. $\bar{P}$ dependence of $P(P_1,P_2)$ of Nus for He and O_2: The meanings of all marks and lines are same as Fig.9.

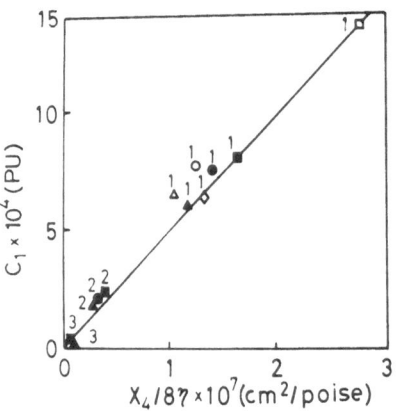

Fig.19. Relationship between C_1 and $X_4/8$:
        1, 2, and 3 are Nu0.8, Nu0.6, and Nu0.1, respectively;
        ●, N_2; o, He; ▲, O_2; △, Ar; ■, H_2; □, CO_2; ◇, air; ──,
        theoretical value of C_1 (eq.(51)).

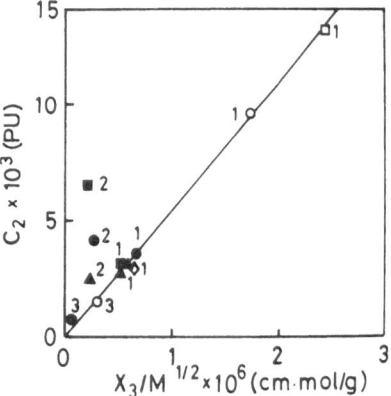

Fig.20. Relationship between C_2 and $X_3/M^{1/2}$:
        1, 2, and 3 mean Nu0.8, Nu0.6, and Nu0.1, respectively;
        ●, N_2; o, He; ▲, O_2; △, Ar; ■, H_2; □, CO_2; ◇, air; ──,
        theoretical value of C_2 (eq.(52)).

in the figure is the theoretical line calculated from eq.(36), in
which F flow alone was taking into account. The excellent agreement of the experimental data with the theory indicates that in case
F F flow alone occurs and $P(P_1,P_2)$ is equivalent to P_F. In Figs.12
- 16, $P(P_1,P_1)$ gradualy increases with P_1 and accordingly, $P(P_1,P_1)$
$P(P_1,0)$ holds. This implies an increase in $\bar{P}$ leads to lower value
of f_0. The pressure dependence of f_0 is almost negligible for
Nu0.03, Nu0.015, Nu0.00 and Nus.

5.2.3 Case D: When the gas flow through porous membrane is VF flow alone

As is evident from Fig.4, VF flow appears in case D without coexisting any other flow, in case B with V flow and in case E with
F flow. The experimental $P(P_1,P_2)$ - $\bar{P}$ relations, corresponding to
case D (designated by closed triangle in the figures) reveals noti-
ceably large slope. $\bar{P}$ dependence of $P(P_1,P_2)$ (i.e., the slope of
the plot) can be evaluated by using eq.(34), in which P_{VF}' is given
by eq.(14) for case D. As a result, $\bar{P}$ dependence of $P(P_1,P_2)$ in
case D should be accompanied with the pressure dependence of (1) f_1,
(2) f_0, (3) x_0 (eq.(46)) and (4) P_H'. Detailed analysis of the experimental dependence of $P(P_1,P_2)$ in case D of the pressure according to eqs.(34) and (14) is tremendously difficult. For the sake
of simplicity, the theoretical $P(P_1,P_2)$, $P(P_1,P_1)$ and $P(P_1,0)$ values
were compared directly with the experimental ones.

Fig.22 is an example, in which He/Nu0.8 and CO_2/Nu0.08 are used.
The experiments cover from case F to case B (as $\bar{P}$ increases). At
first, P_H' was calculated by putting $\eta=1.985 \times 10^{-4}$ poise (in the case
of He) into eq.(2) and P_{Sp}' was estimated by putting $f_1=0.87-5/(P_1+$

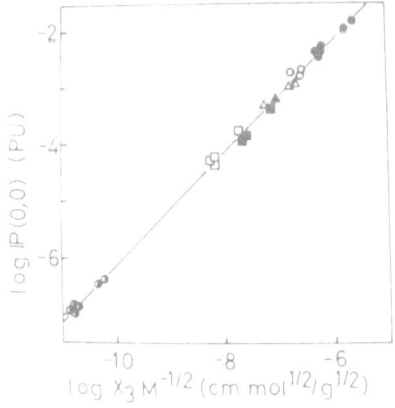

Fig.21.   Log-log plot of $P(0,0)$ against $X_3 M^{-1/2}$:
          ●, Nu0.8; o, Nu0.6; ▲, Nu0.2; △, Nu0.1; ■, Nu0.08; □, Nu0.05
          ◐, Nu0.03, ◑, Nu0.015.

+18), M=4.0 (He), and T=298.15(k) into eq.(3). P_F' was calculated
by assuming $f_o=1.0$ from eq.(4). The P_{VF}' was obtained by taking
the above P_H', P_{Sp}' and P_F' values into eqs(15) and (16). $P(P_1,P_2)$
was counted by using these P values together with experimental $N(r)$
from eq.(26). In the similar manner, $P(P_1,P_1)$ and $P(P_1,0)$ were evaluated from eqs.(24) and (36), respectively. $P(P_1,P_2)$ data belonging to case D (filled triangle in Fig.22) agree well with the theoretical curve at $P_1=70$ cmHg. All experimental $P(P_1,P_1)$ points, belonging to cases B - F, fall well on the theoretical $\bar{P}(P_1,P_1)$ curve.
The $P(P_1,0)$ curve can explain the experimental results of case D.
These facts support the validity of eq.(34) to express case D.

In summary, the permeability coefficient $P(P_1,P_2)$ of the porous polymeric membrane in the case when V or F or VF flow alone
occurs can be most successfuly evaluated from some simplified forms
of eq.(23). The applicability of eq.(23) to cases B, C, E can be
anticipated if the overall permeability coefficient $P(P_1,P_2)$ for the
gas flow with more than two different mechanisms can be represented
by the simple summation of $P(P_1,P_2)$.

5.2.4 Case C with $P_2{\rightarrow}P_1$: When the gas flow through porous membrane consists of V flow and F flow

At the limit $P_2{\rightarrow}P_1$ of case C VF flow disappears, remaining V
and F flows. This corresponds to $P(P_1,P_1)$ (broken line), extrapolated from case C data (open triangle) to $P_2{\rightarrow}P_1$. Eq.(31) is theoretical relation, derived by assuming additivity of the permeability
coefficients obtained for different flow mechanisms. Note that the
porous membranes having the same X_4 may have different α_1 if they

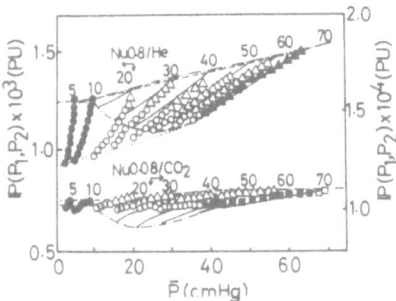

Fig.22. $\bar{P}$ dependence of $P(P_1,P_1)$ of Nu0.8 for He and Nu0.08 for CO_2:
———, —.—., and —..—, calculated curve of $P(P_1,P_2)$, $P(P_1,0)$,
and $P(P_1,P_1)$, respectively, by putting $T_s=273.15K$, $P_s=76$
cmHg, T=298.15K, $f_o=1.0$, $f_1=0.87-5/(P_1+18)$, $N(r)$ given in
Fig.5 and M=4.0, $\eta=1.985\times10^{-4}$ poise for He and M=44.0, $\eta=$
1.38×10^{-4} poise for CO_2 into eqs. (2), (3), (4), (15), (16),
and (26) ($P(P_1,P_2)$) or (24) ($P(P_1,P_1)$) or (36) ($P(P_1,0)$).

have different $N(r)$ and may have different k_1 if they have different r_{max} (eqs.(32) and (33)). For the same X_4, as an increase in k_1 the P_1 dependence of $P(P_1,P_1)$, due to the viscous flow (the first term of the right-hand side of eq.(31)), increases. A large α_1, indicates large proportion of F flow in $P(P_1,P_2)$.

Fig.23 shows the P_1 dependence of α_1 (eq.(32)) and $[1 - \alpha_1/\{ 1+ (1 - \alpha_1)k_1\}]$ (see eq.(33)) for Nu0.8/N_2 calculated by using eqs. (32) and (33). The data points at $P_1=20 - 40$ cmHg belong to case C, and closed circle is $P(P_1,P_1)$. The double dotted chain line in Fig.9(c) is $P(P_1,P_1)$ calculated by putting α_1, k_1 obtained in Fig. 23 and $f_0=1$ and f_1 given by eq.(53) into eq.(31). The agreement between the theory and the experiment is truly remarkable. This ascertains validity of the additivity hypothesis. In this manner the wide applicability of eq.(26) was verified experimentally.

5.2.5 Pressure dependence of reflection coefficients f_0 and f_1

In the previous section, f_1 was determined in the manner that $P(P_1,P_1)$ data in case A coincides with the calculated value by eq. (28). f_1 thus determined is of couse P_1 dependent (designated as $f_1(P_1,P_1)$ as demonstrated for Nu0.8 in Fig.24. f_1 (70,70) values for various inorganic gases/Nu0.8, as collected in Table 2, are found to be 0.84 ± 0.03, being similar to the literature data[9] $0.8 - 0.9$. $f_1(P_1,P_2)$ in Fig.24 can be represented by eq.(53).

$P(P_1,P_2)$ data in Fig.12(a), corresponding to case F, depends notably on the pressure. From eq.(36) this pressure dependence may be explained by the pressure dependence of f_0 $(f_0(P_1,P_2))$, associated with the variation of the amount of gas molecules absorbed on the pore wall with P_1 and P_2.

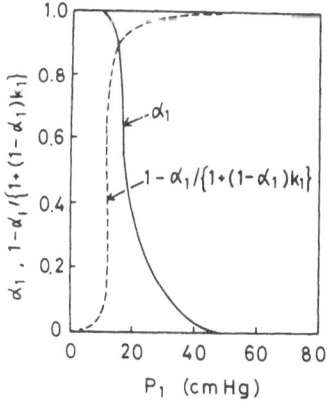

Fig.23. P_1 dependence of α_1 and $[1 - \alpha_1/\{1 + (1 - \alpha_1)k_1\}]$ of poly carbonate membrane Nu0.8 for N_2:
——, α_1; ---, $[1 - \alpha_1/\{1 + (1 - \alpha_1)K_1\}]$.

$f_o(P_1,P_1)$ was estimated from $\mathbb{P}(P_1,P_1)$ data and $f_o(P_1,0)$ from $\mathbb{P}(P_1,0)$ data for He/Nu0.1 in Fig.12(a). Fig.25 shows the pressure dependence of $f_o(P_1,P_1)$ and $f_o(P_1,0)$ thus obtained as open and closed circles, respectively. Both f_o decreases with increasing P_1, but the P_1 dependence of $f_o(P_1,P_1)$ is larger than that of $f_1(P_1,0)$ and semi-empirically expressed by:

$$f_o(P_1,P_1) = 1 - \mathcal{E}P_1 \tag{54}$$

$$f_o(P_1,0) = \mathcal{E}_{00} - \mathcal{E}_0P_1 \tag{55}$$

where $\mathcal{E}$, $\mathcal{E}_0$, $\mathcal{E}_{00}$ are P_1-independent constants for a given combination of gas/polymeric membrane. For example, for He/Nu0.1 $\mathcal{E}=$ 5.3×10^{-3} cmHg^{-1}, $\mathcal{E}_0 = 7.0 \times 10^{-4}$ cmHg^{-1} and $\mathcal{E}_{00} = 1.045$ were evaluated. The difference between $f_o(P_1,P_1)$ and $f_o(P_1,0)$ may reflect the difference of the absorbed state of the gas on the pore wall. $f_o(P_1, 0)$ obtained by method 1 measurement (Fig.6) is f_o of the non-absorbed state, in which gas molecules, collided against the pore wall, exchange its kinetic energy with the wall and move in the hole according to the cosine law. In this case, f_o is 1, by its definition.

In Fig.10(b) the experimental $P(P_1,0)$ value obtained for Nu0.6/ N_2 by methods 1 (••••) and 2 (chain line) are in good agreement with each other at $P_1 > 10$ cmHg. In contrast, at $P_1 \lesssim 5$ cmHg $P(P_1,0)$ by method 2 is considerably larger than $\mathbb{P}(P_1,0)$ by method 1. The difference is order of P_S (for example, $P_S = 1 \times 10^{-4}$ (PU) at $P_1 = 2$ cmHg for Nu0.6). With an increase in time of gas flow t, P_2 increases in method 1, but decreases in method 2. In the former, the absorption equilibrium is not attained and the contribution of P_S to $P(P_1, 0)$ can be ignored, but in method 2 this contribution can never be neglected. When a gas (in particular organic gas) penetrates through the porous membrane with small pore size ($2r_1 \lesssim 35$ nm) under lower P_1, the absorption equilibrium plays a significant role.

5.3 Permeation of Organic Gas through Porous Polycarbonate Membrane

There is no difference in the permeation characteristics in the Adzumi's data on the gas flow in a glass capillary between inorganic and organic gas[13]. For the polymer membranes having large surface area per unit volume as employed in this study the chemical structure of the permeating gas is expected to be possible to affect the $\mathbb{P}(P_1,P_2)$ value. In fact $\mathbb{P}(P_1,P_2)$ of the polymeric membrane with extremely small pore size (for example, $2r=50$ Å) is significantly dependent of the chemical structure of the permeating gas and in addition the separation coefficient α differs greatly from α_o (= $(M_2/M_1)^{1/2}$)[18]. This phenomena may be interpreted in terms of the surface diffusion flow, which is caused by the diffusional flow of the gas molecule adsorbed on the pore wall.

Table 2. Reflection coefficients f_o and f_1 of various gases for Nu0.08

	He	Ar	H_2	N_2	O_2	CO_2	Air
$f_1(70,70)$	0.83	0.87	0.83	0.84	0.82	0.81	0.87
$f_o(5,5)$	1.0	0.98	1.0	1.0	1.0	0.97	–

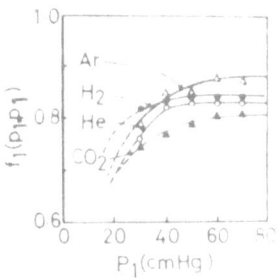

Fig.24. P_1 dependence of $f_1(P_1,P_1)$ of various gases for Nu0.8: ●, H_2 and N_2; o, He; ▲, CO_2; △, Ar.

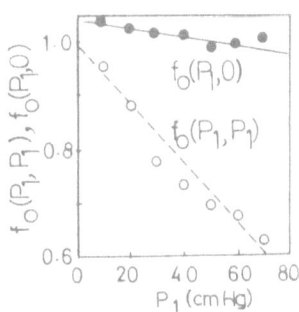

Fig.25. P_1 dependence of $f_o(P_1,P_1)$ and $f_o(P_1,0)$ of He for Nu0.1.

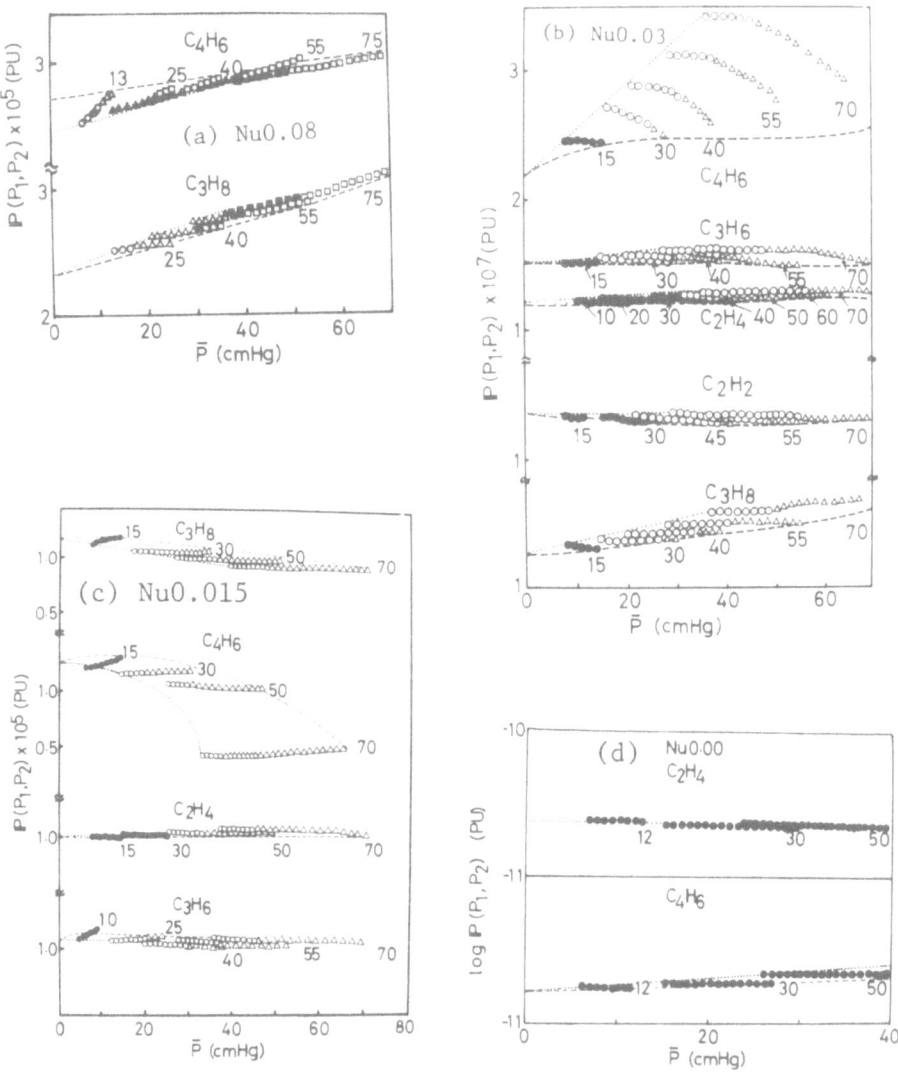

Fig.26. $\bar{P}$ dependence of gas permeability of Nu0.08, Nu0.03, Nu0.015, and Nu0.00 for various organic gases of C_2H_2, C_2H_4, C_3H_8 and C_4H_6:
Numbers on curves denote P_1 (cmHg), all marks and lines have the same meaning as those in Fig.9.

Fig.27 shows the plot of the ratio $(P(0,0 - P_F)/P(0,0)$ as a function of the boiling point for various organic gas for Nu0.03. In the range $T_b \gtrsim -100°C$, the ratio increases rapidly as T_b gets large and in this range $P(P_1,0) \gtrsim P(P_1,P_1)$ holds always, suggesting the necessity of addition of the terms including P_S. The addition term will be discussed in the later section (section 5.5).

Fig.28 shows the difference between experimental $P(0,0)$ and theoretical value of P_F at $P_1=0$ for organic gases as a function of the mean pore size $\bar{r}_3$. The disparity between these two values becomes maximum at $0.8 < \bar{r}_3 < 20$nm.

5.4 Separation of Binary Gas Mixture by Porous Membrane

Fig.29 shows the plot of $P(P_1,0)$ against P_1 for O_2 or CO_2 for various polycarbonate membranes. For Nu0.1 $P(P_1,0)$ of O_2 is larger than $P(P_1,0)$ of CO_2, at least in the range $P_1=0 - 70$ cmHg. For Nu0.8, $P(P_1,0)$ of CO_2 is larger than that of O_2 in the range $P_1 \lesssim 15$ cmHg and $P_1 > 15$ cmHg in reverse.

Fig.30 shows the relations between P_1 and the average pore size of the membrane $(\bar{r}_3 \cdot \bar{r}_4)^{1/2}$, which satisfy the conditions $P(P_1,0)_{O_2} < P(P_1,0)_{CO_2}$. $(\bar{r}_3 \cdot \bar{r}_4)^{1/2}$ is readily evaluated from the water filtration rate method and in addition, $P(P_1,P_2)$ depends strongly on X_3 and X_4. From theoretical point of view, $\bar{r}_4$ should be used in place of $(\bar{r}_3 \cdot \bar{r}_4)^{1/2}$. In the figure, F, Sp and H mean the regions, in which F, Sp and H flows are predominant. The each component will be separable from O_2/CO_2 mixed gases by the gas permeation through the porous polymer membrane when F or (F+Sp) flow predominate. The differences in gas permeation velocity can be attributed to their molecular weight differences.

Table 3 summarizes the experimental results of the mixed inorganic gas penetration through Nu0.1 and Nu0.03. In this case three combinations, He/N$_2$, He/Ar and N$_2$/O$_2$ were employed. In Table 3 is included the separation coefficient α, defined by eq.(38) and the separation   efficiency γ, defined by the relation

$$\gamma = (\alpha - 1)/(\alpha_o - 1) \tag{56}$$

where α_o is the theoretical value of F flow, the separation coefficient calculated by assuming that F flow occurs ideally. At the limit of $P_2 \to 0$ and $P_1 \to 0$, γ approaches to 1.0. Table 3 indicates that the gas mixture can be separated by the gas permeation method as anticipated.

When the gas mixture contains organic gas as a component, γ value inclines to be negative or more than 1.0 as shown in Table 4 for Nu0.03. This indicates that the other mechanisms than F flow govern the organic gas flow for Nu0.03. Anyhow, organic gas mixture can also be separable by using gas permeation of porous membrane.

5.5   Characteristic Features of Surface Diffusion S Flow

Inspection of Figs.27 and 28 suggests strongly that the devi-ation of $P(0,0)$ from P_F can not be ignored when organic gas permea-tes through membrane with $\bar{r}_3$ in the range 0.8 - 20 nm.   Since P_D is very small in this range and $P(0,0)$ value changes depending on the absorption condition of gas on the porous membrane, such as virgin surface or not, the above deviation is originated by the contribu-tion of P_S to $P(0,0)$.   Then, we obtain empirically the following equation for P_S of C_4H_6 for polycarbonate membrane.

$$P_S = P(0,0) - P_F = 2.66 \times 10^{-4} X_1 \quad \text{(PU)} \tag{57}$$

P_S is proportional to X_1, accordingly to the pore surface area.   Fig. 32 demonstrates the r_3 dependence of X_1/X_3 and $(1 - Pr)/X_3$ for Nuc-lepore.   The former represents a measure of the ratio P_S/P_F and the later is that of P_D/P_F (see eqs.(2) - (6)).   The smaller r_3 becomes too small such as less than 10 nm, P_D becomes larger than P_S.   Con-sequently, the pore size range where P_S governs the permeation of gas is very limited.

Fig.33 plots the apparent activation energy Ep (eq.(47)) as a function of the 3rd momentum of $N(r)$ (X_3) for O_2 or C_4H_6/polycarbo-nate membrane.   Ep=-0.4 Kcal/mol in the range $X_3 > 10^{-8}$ cm indicates that both O_2 and C_4H_6 show F flow only and in the range $X_3 < 10^{-14}$cm, D flow is predominant for O_2/polycarbonate and C_4H_6/polycarbonate. At $X_3 \simeq 10^{-13}$ cm S flow predominates for C_4H_6 and (F+D) flow for O_2. Ep of S flow ranges between 0 and 0.4 kcal/mol. which is by far lar-ger than those of F and F flows (ca.-0.4Kcal/mol.).

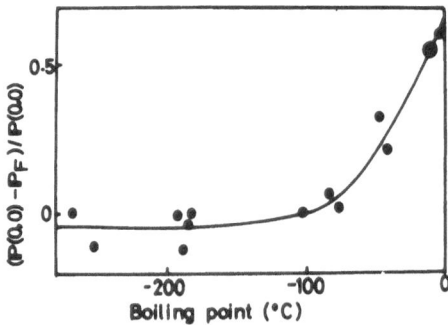

Fig.27.   Boiling point dependence of $(P(0,0)-P_F)/P(0,0)$ of Nu0.03 for various gases:
   ●, organic gas; o, inorganic gas.

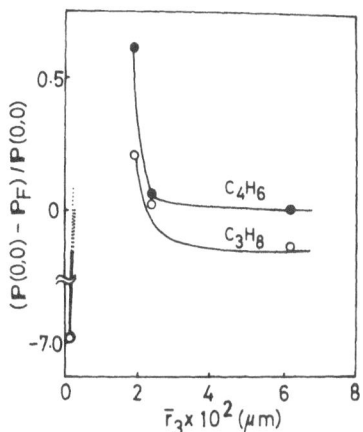

Fig.28. $\bar{r}_3$ dependence of $\{P(0,0) - P_F\}/P(0,0)$.

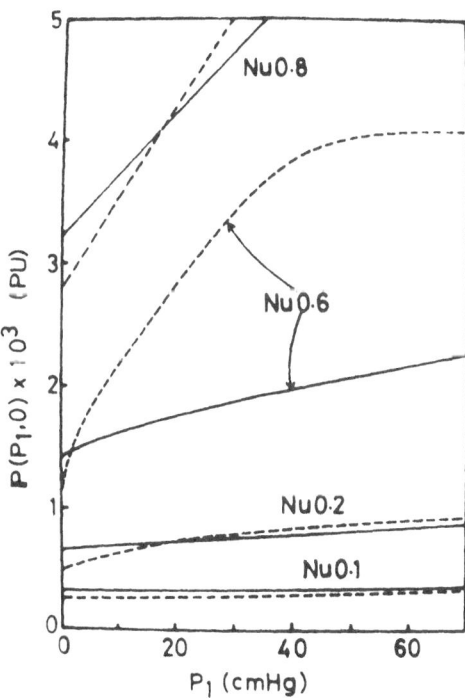

Fig.29. P_1 dependence of $P(P_1,0)$ of polycarbonate membranes Nu0.8, Nu0.6, Nu0.2, and Nu0.1 : full line, CO_2; broken line, O_2.

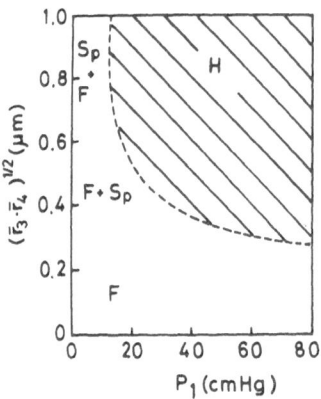

Fig. 30.   Correlation between $(\bar{r}_3 \cdot \bar{r}_4)^{1/2}$ and P_1 which satisfies the condition, $\mathbb{P}(P_1,0)$ for CO_2 $\mathbb{P}(P_1,0)$ for O_2: The hatched region means the region of $\mathbb{P}(P_1,0)$ for CO_2 $\mathbb{P}(P_1,0)$ for O_2.

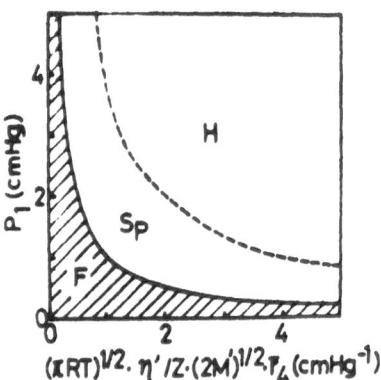

Fig. 31.   Variation of gas permeation mechanisms for binary gaseous mixture with experimental conditions such as P_1 and r_4: H, viscous flow; Sp, slip flow; F, free molecular flow; $\eta'/M'^{1/2}$, the ratio of viscosity to root of molecular weight of mixture of two gases; hatched region denotes the region where gas separation occurs.

Table 3. Change in composition of mixed inorganic gas before and after permeation

Membrane	Component A/B	Experimental pressure (cm Hg) P_1	P_2	Composition of gas $X_1/(1-X_1)$	$X_2/(1-X_2)$	α	γ
Nu 0.01	He/N$_2$	38.0	1.0	0.937	1.327	1.42	0.253
Nu 0.01	He/N$_2$	34.0	0.4	0.888	1.811	2.04	0.631
Nu 0.03	He/Ar	5.17	0.92	1.03	1.82	1.77	0.368
Nu 0.03	He/Ar	44.5	1.05	1.03	2.27	2.20	0.554
Nu 0.03	N$_2$/O$_2$	14.9	0.47	3.73	3.99	1.07	1.01
Nu 0.03	N$_2$/O$_2$	45.1	12.5	3.73	3.95	1.06	0.87

X_1 and X_2 are mol fraction of the gas with smaller molecular weight (component A) of high pressure side and low pressure side, respectively.

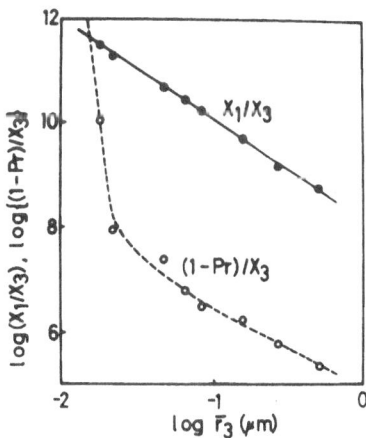

Fig.32. Plots of X_1/X_3 and $(1 - Pr)/X_3$ against $\bar{r}_3$.

Table 4. Change in composition of organic gas mixture before and
after permeation

Membrane	Experimental component A/B	pressure (cm Hg) P_1	P_2	Composition of gas $X/(1-X_1)$	$X_2/(1-X_2)$	α	γ
Nu 0.03	C_2H_4/C_4H_6	9.65	0.63	1.20	0.616	0.513	-1.25
Nu 0.03	C_2H_4/C_4H_6	45.5	2.40	1.20	0.378	0.315	-1.76
Nu 0.03	C_2H_6/C_4H_{10}	10.1	1.68	1.14	1.23	1.08	2.19
Nu 0.03	C_4H_6/C_4H_{10}	62.1	1.86	1.14	1.25	1.10	2.74
Nu 0.03	He/C_4H_6	9.65	1.97	1.03	1.29	1.25	0.093
Nu 0.03	He/C_4H_6	60.2	6.91	1.03	0.587	0.570	-0.16
Nu 0.03	Ar/C_4H_6	10.2	0.50	1.08	0.465	0.431	-2.76
Nu 0.03	Ar/C_4H_6	60.5	8.39	1.08	0.241	0.223	-3.77

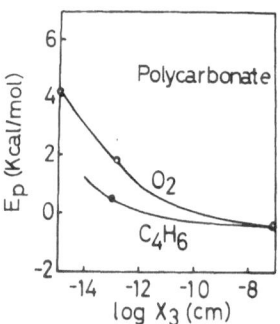

Fig.33. E_p plotted as log X_3 for O_2 or C_4H_6/polycarbonate membrane.

6 CONCLUSION

The gas permeation mechanisms are summarized as Fig.34. The broken circle indicates the mechanisms speculated from the ratio of X_1/X_3 and $(1 - Pr)/X_3$. Table 5 compiles the characteristic features of gas flows through polymeric membrane.

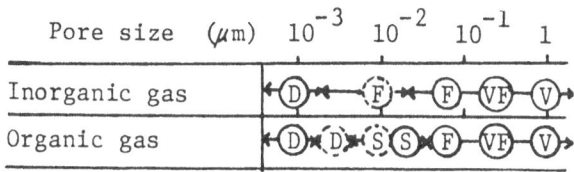

Fig.34. Gas permeation mechanisms.

Table 5. Characteristic features of various gas flows

FLOW	PRESSURE DEPENDENCE P	P_2	VISCOSITY DEPENDENCE	M.W. DEPENDENCE	X_1 DEPENDENCE	ΔH_A (KCAL/MOL)	SEPARATION OF GAS
V	+	+	$1/\eta$	---	X_4	<-0.4	---
F	---	---	---	$M^{-1/2}$	X_3	-0.4	M.W.
SD + F	- (+)	$0\sim-$	---	?	X_1	$0\sim0.4$	MOL. INTERACTION
D	---	---	---	---	$1-2\pi X_2$	>0	DITTO

Acknowledgement

This study was in part financially supported by the Scientific Research Funds of the Ministry of Education of Japan.

REFERENCES

1) See, for example, T. Nakagawa, "Polymers and Water" (e, by The Society of Polymer Sci., Japan), Saiwai, (1972) or V.T.Stannett, WJ. Koros, D.R.Paul, H.K. Lonsdale, and R.W. Baker, Adv. Polym. Sci., 32, 69 (1979)
2) See, for example, H. Adzumi, Bull. Chem. Soc., Japan, 12, 199 (1937)
3) See, for example, E.A. Flood, R.H. Tomlinson, A.E. Leger, Can. J. Chem., 30, 348 (1952)
4) K. Knudsen, Ann. Physik (IV), 28, 75 (1909)

5) T.L. Hill, J. Chem. Phys., 25, 730 (1956)

6) H. Yasuda and J.T. Tsai, J. Appl. Polym. Sci., 18, 805 (1974)

7) I. Cabasso, K.q. Robert, E. Klein, J.M. Smith, J. Appl. Polym. Sci., 21, 1883 (1977)

8) See, for example, P. Millikan, Phys. Rev., 21, 217 (1923)

9) P.C. Carman, Proc. Roy. Soc., 203A, 55 (1950)

10) G.W. Sears, J. Chem. Phys., 22, 1552 (1954)

11) T.L. Hill, J. Chem. Phys., 25, 730 (1956)

12) G.M. Barrow, "Physical Chemistry", McGraw Hill (1973)

13) T. Nohmi, S. Manabe, K. Kamide, T. Kawai, Kobunshi Ronbunshu, 34, 729 (1977)

14) H. Adzumi, Bull. Chem. Soc. Japan, 12, 285 (1937)

15) K. Kamide and S. Manabe, "Ultrafiltration Membranes and Applications" ACS Meeting, Sept., 1979, Washington

16) H. Adzumi, Bull. Chem. Soc. Japan, 12, 292(1937)

17) S.A. Stern, T.F. Sinclair, P.J. Gareis, N.P. Vohldieck, and P. H. Mohr, Ind. Eng. Chem., 57, 49 (1965)

COMPOSITE HOLLOW FIBER MEMBRANES FOR GAS SEPARATION:

THE RESISTANCE MODEL APPROACH

Jay M. S. Henis and Mary K. Tripodi

Corporate Research and Development
Monsanto Company
St. Louis, Missouri   63166

Of the potential limitations to the use of membranes for gas separation, one of the most difficult to overcome has been that of inadequate permeation rate. Indeed, this limitation applies not only to gas separations, but to all separations for which membranes have been used.

A major advance in the field of membrane separations which dealt with this separation was made by Loeb and Sourirajan[1]. By casting a solution of cellulose acetate in a specific manner, an asymmetric film was produced which was used for desalination. This film had a relatively thin, dense surface layer. Equation 1 is a general expression relating flux Q to the variables P, Δc, A and ℓ (i.e., permeability constant, concentration gradient, area, and membrane thickness, respectively). By using Equation 1, it can easily be determined that an asymmetric film (total thickness $\sim$100 microns) with a dense outer layer of 1000Å to 10,000Å (typical ot such films) should be 100 to 1000 times faster for permeation than a dense film of similar dimensions made from the same material.

$$Q_i = \frac{P_i \Delta c_i A}{\ell} \qquad \cdots \cdots (1)$$

Since the development of the first asymmetric films[1], it has been found that most polymers (other than cellulose acetate) are relatively difficult to cast or spin into useful asymmetric membranes in film or hollow fiber form. This is particularly true where gas separations are involved. It is often found that asymmetric films and hollow fibers which appear to be useful for liquid separations (including reverse osmosis) are porous to gases

and cannot effect separation. Such pores need be only 5–10Å in
diameter to permit the passage of most permanent gases across the
membrane without actual permeation through the polymer.

Various methods of casting or post-treating asymmetric membranes to eliminate surface pores have been developed for different
polymers. However, the result of such approaches is usually to
densify the surface layer (increasing ℓ substantially) and to decrease substantially the rate achievable through the membrane.

This paper describes a general method by which this limitation
may be overcome. Using the approach described, porous (non-
separating) flat membranes or hollow fibers may be made to separate
by coating the porous substrate membrane with an appropriately
chosen second material (e.g., polymer whose permeability characteristics are matched to the intrinsic permeation properties and the
porosity of the non-separating substrate. The unique feature of
these composite membranes is that the substrate (not the coating)
becomes the effective separating material after the coating has
been applied. A detailed treatment and an empirical model which
explains the observed performance of such a composite membrane has
been reported elsewhere[2]. However, one of the more significant
aspects of such membranes is the very high rates which can be
achieved, and the degree to which the effect of porosity can be
eliminated.

There are actually several advantages which accrue from using
the substrate membrane as the effective separating material. Most
glassy polymers with relatively high permeabilities, which could be
useful as gas separators, are easy to spin as asymmetric hollow
fibers of relatively low surface porosity (less than 0.01% of the
surface available as open pore surface). However, even such relatively low porosity hollow fibers are too porous to be useful for
gas separations.

The resistance model in Reference 1 shows that the use of an
appropriate coating permits a controlled high rate RM composite
hollow fiber membrane to be produced. In such a fiber, occasional
defects or changes in surface porosity will not result in significant changes in fiber performance characteristics. This can be
attributed to the ability of the coating material to plug the
defects or pores in the substrate fiber surface while not adversely
affecting its separating properties. Figure 1 illustrates the
calculated effect of surface porosity on the permeation properties
of a polysulfone hollow fiber and a silicone coated polysulfone
RM composite fiber for a hydrogen/carbon monoxide gas mixture. It
is clearly seen in this Figure that the same variation in fiber
surface porosity is expected to have a much more significant effect
on the permeation properties of an uncoated fiber than those of a
coated fiber. From a practical standpoint, this means that much

less control of, and more variability in, the surface porosity of a fiber can be tolerated without critically affecting the permeation characteristics (rate and separation factor) of the fiber in an RM composite membrane system.

RM composite membranes which separate through the substrate polymer instead of the coating material have two unique advantages over conventional composite membranes made by placing an ultra-thin coating over a highly porous support.

The first advantage is that an ultra-thin coating is not required in an RM composite membrane, as discussed previously. Consequently, the sophisticated coating techniques and procedures for making ultra-thin composite membranes[3] can be avoided. In addition, an ultra-thin coating of a high modulus, glassy polymer would be easily damaged by abrasion or flexing, whereas a relatively thick coating of a rubbery polymer on an RM composite membrane is difficult to damage and actually serves to protect the separating layer of the substrate from damage.

TABLE 1. Comparison of Model Permeation Predictions (Ref. 2) for an Ultra-thin Polysulfone Film on an Impermeable Porous Support Versus a Silicone Coated Polysulfone RM Composite

Membrane	Separating Layer Thickness ($\mathring{A}$) Assumed	$^a(P/\ell)H_2 \times 10^{-6}$	$SF^{H_2}_{CO}$
Ultra-thin film on support[b]	50	240	40
Ultra-thin film on support[b]	500	24	40
RM Composite	1000	98	38
RM Composite	5000	23	38

[a]Units of cc/cm^2-sec-cmHg.

[b]Assume impermeable support with 10% surface porosity and very open substructure (R_4 is negligible).

The second advantage afforded by RM composite membranes is the variability in substrate surface porosities which can be tolerated and still provide a composite membrane with good gas separating properties. This is not the case for a conventional composite membrane. Conventional composite membranes must be formed from substrates which have a high degree of surface porosity (∿10% open) in order to achieve reasonable permeation rates. The porosity of such conventional composite membranes must be controlled closely, not only for number of pores, but also for size distribution, since the coating layer will have to be strong enough to bridge across the open pore.

In Table I the permeation properties of a conventional ultrathin composite membrane of polysulfone on a porous support of an impermeable material (e.g., sintered metal) have been modeled using the resistance model approach and compared to the permeation properties calculated for a typical RM composite membrane of a silicone coated polysulfone asymmetric hollow fiber. It can be seen that, depending upon the degree of porosity in the impermeable substrate of the conventional composite membrane, the ultra-thin coating would have to be a dense layer (with no imperfections) ranging in thickness from 50 to 500Å to yield permeation results similar to those which are easily achieved with RM composite membranes of polysulfone with silicone coatings ranging in separating layer thickness from 1000 to 5000Å.

REFERENCES

1. S. Loeb and S. Sourirajan, Report No. 60-60, Department of Engineering, University of California, Los Angeles, 1960.

2. J. M. S. Henis, M. K. Tripodi, J. Membrane Science (to be published).

3. W. J. Ward III, W. R. Browall, and R. M. Salamme, J. Memb. Sci. 1, 99 (1976).

PURIFICATION OF ENZYMES BY AFFINITY CHROMATOGRAPHY

Shigeo Katoh, Masami Shiozawa and Eizo Sada

Chemical Engineering Department
Kyoto University
Kyoto 606, Japan

INTRODUCTION

Affinity chromatography has provided a new method for the purification of biological materials. The method depends on the highly
specific interactions between pairs of biological materials, such as
enzyme-substrate, enzyme-inhibitor and antigen-antibody. One of the
interacting components(the ligand) is immobilized onto an insoluble
support, and the other component is selectively adsorbed onto the
ligand. The adsorbed component can then be eluted with a solution
which weakens the interactions between the two components.

In the present work trypsin and β-galactosidase were purified
with several kinds of ligands coupled to agarose beads(Sepharose 4B
and 6B). The effects of the affinity between the ligands and the
enzymes, the concentration of buffer solutions and the molecular
weight of the enzymes on the amount and the purity of the eluted
enzymes were studied. The selective elution of trypsin with an
inhibitor(benzamidine) was also compared with nonselective elution.

EXPERIMENTAL

Bovine trypsin(crystallized twice), β-galactosidase(from *E.
coli*), soybean trypsin inhibitor(STI), p-aminobenzamidine(BA),
p-aminophenyl-β-D-thiogalactopyranoside(PAPTG), o-nitrophenyl-β-D-
galactopyranoside, benzoyl arginine p-nitroanilide HCl and
p-nitrophenyl p'-guanidinobenzoate HCl were obtained from Sigma Chemicals Co.

The adsorbents used for the purification of trypsin were CNBr-

activated Sepharose 4B-STI, CH-Sepharose 4B-STI, epoxy-activated
Sepharose 6B-STI and CH-Sepharose 4B-BA. β-Galactosidase was purified with Sepharose 4B-PAPTG and epoxy-activated Sepharose 6B-PAPTG.
Details of the preparation of the adsorbents are shown in the previous papers[1,2]. The adsorbents, buffer solutions and nonselective
eluents used in this work are shown in Table 1.

Figure 1 is a schematic diagram of the experimental apparatus.
The transparent acrylic resin column of 1.7 cm i.d. was equipped
with a water jacket and kept at 10°C. At the bottom of the column
was placed a fritted glass filter plate to prevent effusion of adsorbent. The height of the adsorbent bed was varied from 5 cm to
10 cm. The column was equilibrated with a buffer solution. The
enzyme dissolved in the buffer solution was pumped into the column,
and when the enzyme concentration of the effluent solution reached
a predetermined value, the column was washed with the same buffer
without the enzyme. Then the enzyme was eluted with the eluent.
The effluent solution, the absorbance of which was continuously
measured with a spectrophotometer(Shimadzu, Spectronic 70-UV), was
collected at scheduled intervals with a fraction collector. After
elution the column was washed with 8 M urea solution and reequili-
brated with the buffer solution. The activity of the enzyme was
determined from the hydrolysis of synthetic substrates(benzoyl
arginine p-nitroanilide HCl for trypsin[3] and o-nitrophenyl-β-D-
galactopyranoside for β-galactosidase[4]). The amount of active
trypsin was measured by active site titration with p-nitrophenyl
p'-guanidinobenzoate HCl[5] and the purity of trypsin was determined
from the amount of active trypsin and the optical factors for
trypsin[6] and contaminating proteins.

Table 1. Adsorbents and buffer solutions

adsorbent	adsorbed component	buffer	eluent
Sepharose 4B-STI CH-Sepharose 4B-STI CH-Sepharose 4B-BA	trypsin (MW = 24000)	Tris-HCl + CaCl$_2$ (pH 7.8)	0.02N HCl + 0.5M NaCl
epoxy-activated Sepharose 6B-STI	trypsin	0.05M Tris-HCl + 0.02M CaCl$_2$ + 0.5M NaCl (pH 7.8)	0.02N HCl + 0.5M NaCl
CH-Sepharose 4B-PAPTG	β-galactosidase (MW = 540000)	0.035M Tris-HCl (pH 7.4)	0.05M borate buffer (pH 11)
epoxy-activated Sepharose 6B-PAPTG	β-galactosidase	0.05M Tris-HCl + 0.1M NaCl (pH 7.5)	0.05M borate buffer (pH 11)

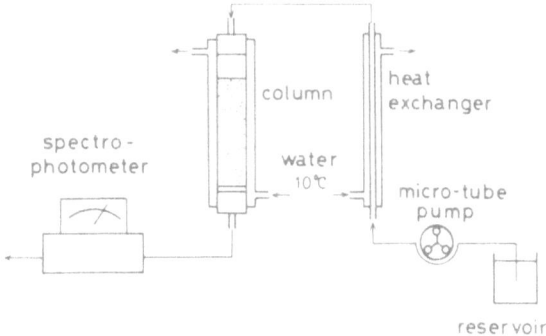

Fig. 1.   Experimental apparatus.

Trypsin adsorbed onto Sepharose 4B-BA was selectively eluted with 4 mg/cm^3 of benzamidine dissolved in the 10^{-3} M Tris buffer, pH 7.8.   The pH of the eluted solution was adjusted to 1.2, at which value the interaction between trypsin and benzamidine weakened. Trypsin was then separated from benzamidine by gel chromatography with Sephadex G 15.

RESULTS AND DISCUSSION

Figure 2 shows a trypsin adsorption and elution profile for the CH-Sepharose 4B-BA column.   Contaminating proteins emerged first from the column, as shown by the fisrt, flat portion of the adsorption profile, followed by trypsin.   When c/c_0 reached 0.05, the column was washed with the same buffer solution without trypsin. Finally, bound trypsin was eluted.

Table 2 shows the amounts and the purity of trypsin eluted from the CH-Sepharose 4B-STI column with buffer solutions of varying concentration.   Above the buffer concentration of 0.05 M Tris, the amount of trypsin eluted was not influenced by the concentration of the buffer solution because of the high affinity between trypsin and STI.   Since nonselective adsorption of impurities on agarose beads is suppressed at high buffer concentrations, the purity of the eluted trypsin was almost 100 %.

With the CH-Sepharose 4B-BA column, as shown in Table 3, the amount of trypsin nonselectively eluted decreased with the increase in concentration of the buffer solution because of the relatively low affinity of trypsin for this adsorbent.   At lower concentrations

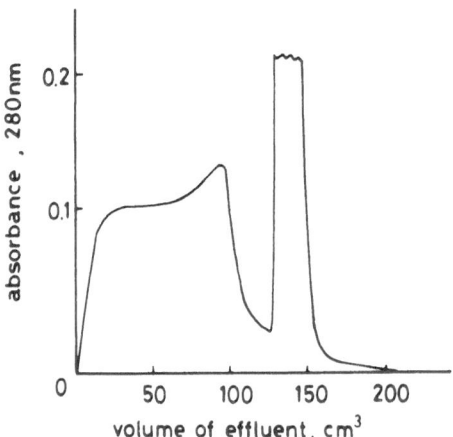

Fig. 2.   Adsorption and elution profiles.   (CH–Sepharose 4B–BA,
          liquid flow rate 0.041 cm^3/sec).

Table 2.   Trypsin purification by CH–Sepharose 4B–STI
           (purity of crude trypsin   62.3 %).

buffer solution (pH 7.8)	complete saturation washing for 15 min		stopped at $c/c_o = 0.05$ washing for 15 min	
	amount mg trypsin/cm^3-bed	purity,%	amount mg trypsin/cm^3-bed	purity,%
0.01M Tris-HCl + 0.004M CaCl$_2$	4.84	100		
0.05M Tris-HCl + 0.02M CaCl$_2$	2.11	97		
0.05M Tris-HCl + 0.02M CaCl$_2$ + 0.5M NaCl	1.90	93	2.05	97
0.05M Tris-HCl + 0.02M CaCl$_2$ + 2.0M NaCl	2.25	100		

Table 3. Trypsin purification by CH-Sepharose 4B-BA, (purity of crude trypsin 62.3 %).

buffer solution (pH 7.8)	complete saturation washing for 15 min		stopped at c/c_0 = 0.05					
			washing for 5 min		washing for 15 min		washing for 15 min eluted by BA	
	amount mg trypsin/cm^3-bed	purity, %	amount mg trypsin/cm^3-bed	purity, %	amount mg trypsin/cm^3-bed	purity, %	amount mg trypsin/cm^3-bed	purity, %
10^{-3}M Tris-HCl + 4 × 10^{-4}M CaCl$_2$	8.98	93	8.39	77	7.85	80	7.38	94
2.5 × 10^{-3}M Tris HCl + 10^{-3}M CaCl$_2$	3.02	93	2.51	87				
10^{-2}M Tris-HCl + 4 × 10^{-3}M CaCl$_2$	0.93	100	1.03	81	0.78	100		

of the buffer solution the nonselective adsorption of impurities
increased. However, when trypsin was eluted after complete saturation of the column, its purity was almost 100 %. This was because
impurities temporarily adsorbed on the ligand were replaced by
trypsin before complete saturation was reached. On the other hand,
the purity of eluted trypsin fell with the decrease of the buffer
concentration when the supply of trypsin solution was stopped at
c/c_0 = 0.05. Washing with the buffer for 15 min raised the purity
to 100 % at the Tris-HCl concentration of 10^{-2} M but not at that
10^{-3} M.

With CH-Sepharose 4B-STI as adsorbent, therefore, operation at
high buffer concentration will give higher purification, while with
CH-Sepharose 4B-BA it will be necessary to select buffer concentration on the basis of column capacity and the purification required.
An alternative method in this case is selective elution of trypsin
with an inhibitor. When trypsin was selectively eluted for 30 min
with a benzamidine solution of 4 mg/cm^3 in 10^{-3} M Tris buffer, the
eluted amount and purity were as shown in Table 3. The amount of
eluted trypsin was nearly equal to that obtained by nonselective
elution and the purity was almost 100 %. The requirements of amount
and purity are thus met by the selective elution, although an additional process for separation of trypsin and benzamidine is necessary.

Performance of affinity chromatography depends also on the
mass-transfer rate of the adsorbed component into the adsorbent.
Since the length of the adsorption zone in the column incereases as
the mass-transfer rate decreases, the proportion of adsorbent beads
used effectively for adsorption at a break point also decreases.

The mass-transfer rate depends on the molecular weight of the
adsorbed components and the degree of cross-linking of the insoluble
support. Table 4 shows the volumetric coefficients of mass-transfer
evaluated from the break-through curves for trypsin in Sepharose
4B-STI and epoxy-activated Sepharose 6B-STI and for β-galactosidase
in CH-Sepharose 4B-PAPTG and epoxy-activated Sepharose 6B-PAPTG on
the assumptions of a constant pattern of concentration in the
adsorption zone and a linear-driving-force[2]. Sepharose 6B beads
have a higher degree of cross-linking than Sepharose 4B. The mass-
transfer coefficients of trypsin(MW 24000) depend slightly on the
degree of cross-linking of the supports. The coefficients of
β-galactosidase(MW 540000) are lower than those of trypsin and
depend remarkably on the degree of cross-linking.

Column productivities, defined as the amount of product per
operation time per unit cross-sectional area of the column, were
calculated with the following assumptions.
   1) Supply of the adsorbed component is stopped when c/c_0
       reaches 0.1.

Table 4. Volumetric coefficients and column productivities.
(liquid flow rate, 0.045 cm^3/sec)

adsorbent	volumetric coefficient 1/sec	bed height cm	column productivity
Sepharose 4B-STI	0.0147	10	3.28, mg/hr cm^2-column
epoxy-activated Sepharose 6B-STI	0.0135	10	3.26, mg/hr cm^2-column
CH-Sepharose 4B-PAPTG	0.0080	10	2.54, units/hr cm^2-column
epoxy-activated Sepharose 6B-PAPTG	0.0025	20	0.63, units/hr cm^2-column

2) The volume of the buffer solutions needed for washing, elution and reequilibration of the column is 15 times the column volume.

3) The break-through curves obtained from the results of Hashimoto et al.[7] are valid.

The amount of enzyme adsorbed before c/c_0 reaches 0.1 was calculated from the numerical solutions of the break-through curve and these values were divided by the time needed for one cycle to obtain the column productivity. The column productivities calculated with the values of $\rho_b \bar{q}_0/c_0 = 15$ and $c_0 = 0.1$ mg/cm^3 or units/cm^3 are shown in Table 4. The productivities of trypsin are almost equal for the Sepharose 4B and 6B columns. For β-galactosidase the productivity of the Sepharose 6B-PAPTG column is much lower than that of the Sepharose 4B-PAPTG column. Therefore, in the former case the Sepharose 6B column is preferable because of the operational stability of the highly cross-linked beads, and in the latter case the Sepharose 4B column should be used because of the high coefficient of mass-transfer.

CONCLUSION

In the case of a ligand with relatively low affinity with the adsorbed component, the amount of the adsorbed component nonselectively eluted decreases with the increase in concentration of the buffer solution in which it is dissolved, whereas its purity increases. This conflict of requirements for high amount and high purity of the adsorbed component eluted may be circumvented by use of selective elution with inhibitors. Since the performance of affinity chromatography depends also on the mass-transfer rate of the adsorbed component into the adsorbent, the degree of cross-linkage of the support should be selected depending on the molecular weight of the adsorbed component.

NOMENCLATURE

c enzyme concentration in effluent, mg/cm^3

c_0 enzyme concentration at column inlet, mg/cm^3

$\rho_b \bar{q}_0$ adsorption capacity of bed, mg/cm^3

REFERENCES

1. S. Katoh, T. Kambayashi, R. Deguchi, and F. Yoshida, Biotech. Bioeng., 20: 267 (1978).
2. S. Katoh and E. Sada, J. Chem. Eng. Japan. 13: 151 (1980).
3. B. F. Erlanger, N. Kokowsky, and W. Cohen, Arch. Biochem. Biophys., 95: 271 (1961).
4. E. Steers, Jr., P. Cuatrecasas, and H. B. Pollard, J. Biol. Chem., 246: 196 (1971).
5. T. Chase, Jr. and E. Shaw, Biochem. Biophys. Res. Comm., 29: 508 (1967).
6. F. C. Wu and M. Laskowsky, J. Biol. Chem., 213: 609 (1955).
7. K. Hashimoto, K. Miura, and M. Tsukano, J. Chem. Eng. Japan, 10: 27 (1977).

HYDROPHOBIC AND OTHER NON-IONIC PARAMETERS IN PROTEIN

SEPARATION AND ADSORPTIVE IMMOBILIZATION BY SUBSTITUTED AGAROSES

B.H.J. Hofstee

Biochemistry Division
Palo Alto Medical Research Foundation
Palo Alto, California 94301

Studies on the interaction of proteins with rather insoluble (apolar) compounds can be made in an aqueous environment by attaching such compounds covalently to a hydrophilic (wettable) matrix such as is provided by porous glass, polyacrylamide, cellulose or agarose (for comparison of the properties of these materials see ref. 1). For the present investigations various n-alkylamines or ω-phenyl-n-alkylamines were attached to agarose (Sepharose CL-4B) via CNBr-activation of the polysaccharide (2). The agarose-bound amines retain their basic properties and are positively charged under experimental conditions (3). Their relative degrees of substitution were determined from the binding capacities of a negatively charged dye (Ponceau S) in the absence of salt (4). Systematic studies on the binding of purified proteins were made by applying them to small columns of these substituted agaroses equilibrated with solutions of varying composition. The amount of protein bound was determined from the absorbance (A_{280}) of the filtrates. In order to eliminate electrostatic effects on the binding, i.e., to quench the charge on the NH-connector-group of the ligand, the medium usually contained 1-4 M NaCl. The salt at the same time enhances the purely hydrophobic aspects of the interaction (5-7).

Non-Specific versus Specific Binding. It was noted (5) that certain n-alkylamino-agaroses with ligand groups "designed" to have specific affinity for certain enzymes, also were able to bind an array of other proteins unrelated to the enzymes and evidence was presented that this non-specific binding was largely through the hydrophobic "spacer" chain and/or the charged -NH-connector group of the ligand (see also refs. 8,9). Particularly in view of the generality of hydrophobic parameters in protein binding (see below) this factor,

87

in addition to charge effects has been found to be a serious
drawback in the separation and purification of proteins by
"bio-specific affinity chromatography". On the other hand, these
hydrophobic and other non-ionic effects (see below), provide many
new parameters for the chromatographic separation and purification
of proteins as well as compounds of smaller molecular size. As
compared to ion-exchange chromatography, where the choice is limited
to neither positive or negative, non-ionic adsorption chromatography
provides an almost unlimited variety of adsorbents based on "structural" interactions with proteins. The feasibility of protien separ-
artion by differential hydrophobic adsorption - first suggested
by Yon (10) - was further established in a subsequent report from
this laboratory (5). Similar investigations were carried out at
about the same time by others, using the same or similar adsorbents
(6-8, 11, 12).

In order to eliminate the possibility of ionic interaction
altogether, several authors have prepared agaroses with hydrophobic
but uncharged ligands (6,13-15). However, it has been shown (16)
that the results on the binding of several proteins by uncharged,
e.g., n-alkyl-glycidyl-agaroses (13) and by the charged n-alkylamino-
agaroses are essentially the same, provided a sufficient amount of
salt (1-4 M NaCl) is present in the medium to quency charge effects.
Also, in the absence of a convenient procedure for determining the
contents of aliphatic hydrophobic groups directly, the charge on
the -NH-group of the n-alkyl ligands provides a means for estimating
the relative ligand content of these adsorbents (4).

Generality of Hydrophobic Protein Binding. In view of the fact that
the amino acids with hydrophobic side chains generally are located
in the interior of the protein molecule, it was believed until rather
recently that the occurrence of hydrophobic surface groups was rela-
tibely rare and limited to special sites, such as the active center
of certain enzymes. However, it was pointed out by Klotz (17)
that accessible hydrophobic groups occurred much more frequently on
several proteins of known three dimensional structure than had been
assumed. In accord with this finding we noted (18) that almost all
of a large number of proteins chosen at random were bound, although
to greatly varying extents, by the above n-alkylamimo-agaroses. The
binding occurred particularly in the presence of high salt (NaCl)
concentrations, indicating the   absence of charge effects and the
occurrence of (salt stable)hydrophobic binding. It should be noted
that high salt concentrations tend to stabilize the hydrophobic
interior of protein molecules and also decrease the ligand hydrocarbon chain-length needed for binding and thus minimizes protein
denaturation through detergent action (see also below).

Modifying Factors in Hydrophobic Protein Binding. It was found (16)
that certain proteins (e.g., γ-globulin and chymotrypsinogen) are
more extensively bound by agaroses substituted with ω-phenyl-n-
alkylamines as compared to aliphatic n-alkyl-amines of the same or
even lesser hydrophobicity and present in about the same concentration. This was not the case, or even the opposite was true, for

certain other proteins, (e.g., bovine serum albumin and β-lactoglo-
bulin). Other modifying factors found to affect the relative extent
of hydrophobic binding of different proteins were the ligand density
and the composition of the medium, e.g., the salt content. Such
factors provide additional parameters for the chromatographic separation of proteins, but are not conducive to the exact determination
of relative hydrophobicities of proteins from the relative extent
of binding under a fixed set of conditions.

Aromatic-Hydrophobic Effects. Aromatic (π-π) interaction between
adsorbents (e.g., the above mentioned ω-phenyl-n-alkylamino agaroses)
and proteins can be expected to be inherently hydrophobic, at least
in part. However, whereas aromatic binding is through direct interaction (i.e., the sharing of electrons), hydrophobic interaction is
an indirect lyotropic effect, that is enhanced by the presence of
non-chaotropic salts. Since the aromatic effect per se is not
necessarily affected by salt, the relative extent of these two forces
can be manipulated by the type and concentration of salt in the
reaction mixture. Application of these findings has been made to
the selective binding of immunoglobulin by the ω-phenyl-n-alkyl-amino-
agaroses. It was shown (19) that although at extremely high salt
concentrations (e.g., 4 M NaCl) the binding by these adsorbents was
primarily hydrophobic, at lower concentrations (0.5-1.0 M) of the
salt the protein showed a distinct preference for the aromatic ligands
as compared to n-alkylamino-agaroses of about the same hydrophobicity
and ligand content. Furthermore, the binding was found to be inhomogenous and fractionation of the protein could be achieved on the basis
of the ligand hydrocarbon chain length of the adsorbent. These
results possibly bear a relationship to the inherent inhomogeneity
of immunoglobulin, i.e., the presence of various antibodies, and to
the apparently aromatic nature of the hapten binding sites (20).

Non-Ionic Non-Hydrophobic Effects. Aromatic (π-π) interaction is
one of a variety of modes of interaction categorized under the general
term of "charge (electron) transfer" reactions in which also hydrogen bonding (21) and metal-chelate interaction (22) may be included.
Examples of applications of such forces to the differential binding
and purification of materials, varying from low molecular weight peptides to viruses and even whole cells may be found in the proceedings
of the latest International Symposium on Affinity Chromatography (23).

Our experience with non-ionic non-hydrophobic adsorption chromatography is limited to the finding that some proteins, in particular
γ-globulin, are bound by inactivated CNBr-treated agarose (24).
Since the binding occurs at high salt (NaCl) concentration and decreases with the introduction of hydrophobic groups, i.e., the interaction apparently is non-ionic and non-hydrophobic, this binding has
been ascribed to hydrogen bonding, possibly involving carbamide amide
groups that comprise a large fraction of the N-content of the adsorbent (25)

Adsorptive (reversible) Protein Immobilization. This type of protein
binding by substituted solid but hydrophilic matrices is defined as
one that cannot be reversed by continuous washing with the ambient

medium for extended periods of time (days, weeks). It is generally
ascribed to the simultaneous multiple-point interaction of a protein
molecule with several immobilized ligand molecules. Such immobilization may occur through electrostatic interaction, e.g., with clay
particles (26), nucleic acids (27) or other (synthetic) ion exchangers
(28), through hydrophobic interaction (29,30), through charge transfer,
e.g., hydrogen bonding (24), aromatic interaction (31) or through
a combination of two or more of these factors (see below). However,
despite the virtual irreversibility of such combinations the protein
often may be recovered merely by changing the composition of the
medium. When such binding depends on electrostatic interaction
reversal can be obtained simply by raising the ionic strength of
the medium. For the case of hydrophobic interaction this can be
achieved by polarity reducing agents, e.g., amines or alcohols with
hydrophobic groups, or by agents such as ethylene glycol or dimethyl-
formide which possibly have a "chaotropic" effect on the intermolecular hydrogen bonded structure of water (24). When the binding is
through a combination of two or more of these factors a combination
of two or more of such eluants may be needed. However, indiscrimin-
ate application of a protein to an adsorbent, for instance the
application of an extremely hydrophobic protein to an extremely
hydrophobic highly substituted adsorbent may result in adsorption
that cannot be reversed, not even by eluants (e.g., 7 M urea) that
would denature the protein (see ref. 16). In such a case not only
the protein will be lost but also the adsorbent will be rendered
useless for further application.

Protein Separation by Differential Binding on a Hydrophobic Affinity
Gradient. In view of large differences in the affinities of various
proteins for hydrophobic adsorbents (16), it is possible that certain
proteins in an unknown mixture, applied to an arbitrarily chosen adsorbent for the purpose of separation by differential elution, may
be bound with such tenacity that they cannot be recovered without
denaturation, whereas others may not be held at all. In order to
circumvent this possibility it is expedient to lead the protein mixture through a series of interconnected columns of adsorbents of
increasing hydrophobicity, e.g., the above mentioned homologous
series of n-alkylamino-agaroses (32). Each protein in the mixture,
washed into the gradient with ample amounts of the irrigant medium,
tends to be held by the adsorbent that provides the minimum degree
of hydrophobicity required for binding. Even when the proteins are
washed-in with large amounts of the medium and eventually become
virtually immobilized, they usually still can be eluted by a relatively mild eluant (e.g. ethylene glycol) from the subsequently dis-
connected columns, provided the spilling over of protein onto a column
more hydrophobic than required has been avoided by using subsaturing
amounts of protein. A similar procedure for the purpose of avoiding
extremely strong binding, had previously been employed by Porath
for charcoal chromatography (33). It has been found in the meantime
(submitted for publication) that even highly purified proteins most
often are inhomogenous with respect to this type of gradient chroma-

tography. It would appear, therefore, that the procedure is more suitable to the analysis and further fractionation of proteins already highly purified by other means than to the separation of crude mixtures such as tissue extracts or body fluids (see also ref. 16).

Adsorptive Immobilization of Enzymes. The above discussed immobilization of protein by substituted agaroses depends on accessible groups that may occur on the molecular surface of any protein, including enzymes. Thus if an enzyme is applied to such an adsorbent in the presence of its substrate, interaction is likely to occur through groups not located in the enzyme active center, the latter remaining free to convert the substrate into its product(s). Such enzyme "reactors" based on adsorptive immobilization and able to continuously operate for extended periods of time (days, weeks) have been produced on several occasions. Until relatively recently the adsorption was based mainly on electrostatic (ionic) interaction (for a review see ref. 34). However, in view of the above described types of non-ionic multiple point protein binding, based on hydrophobic and/or charge-transfer forces, the range of possibilities for adsorptive immobilization would be greatly increased. Such immobilization could occur through hydrophobic interaction and/or charge transfer, even in the case of a protein with an overall charge the same as that of the adsorbent, at least in the presence of charge quenching salt. Various combinations of non-ionic factors in addition to an overall opposite charge, could provide conditions favorable for the adsorptive immobilization of almost any enzyme (31,35).

ACKNOWLEDGMENTS

This work was supported by U.S. Public Health Service Grants GM 22545 and RR 05313, by Santa Clara County United Way and by the Harvey Bassett Clarke Foundation.

REFERENCES

1. Porath, J. and Kristiansen, T. (1975) in The Proteins (Neurath, H. and Hill, R.L., eds.) 3rd ed. Vol. 1, pp. 95-178, Academic Press, New York.
2. Axen, R., Porath, J. and Ernback, S. (1967) Nature (London) 214, 1302-1304.
3. Porath, J. (1968) Nature 218, 834-838.
4. Hofstee, B.H.J. (1974) in Immobilized Biochemicals and Affinity Chromatography (R.Bruce Dunlap, ed.) pp. 43-59, Plenum, New York.
5. Hofstee, B.H.J. (1973) Anal. Bioch. 52, 403-448.
6. Porath, J., Sundberg, L., Fornstedt, N. and Olsson, I. (1973) Nature 245, 465-466.
7. Hjerten, S. (1973) J. Chromatogr. 87, 325-331.
8. Er-el, Z., Zaidenzaig, Y. and Shaltiel, S. (1972) Biochem. Biophys. Res. Comm. 49, 383-390.

9. O'Carra, P. Barry, S. and Griffin, T. (1973) Biochem. Soc. Trans. 1, 289-290.

10. Yon, R.J. (1972) Biochem. J. 127, 765-767.

11. Rimerman, R.A. and Hatfield, G.W. (1973) Science 182, 1268-1270.

12. Doellgast, G.J., Memoli, V.A., Plaut, A.G. and Fishman, W.H. (1973) Abstr. Fed. Proc. 33(11), 1561.

13. Hjerten, S., Rosengren, J. and Pahlman, S. (1974) J. Chromatogr. 101, 281-288.

14. Jost, R., Miron, T. and Wilchek, M. (1974) Biochim. Biophys. Acta 362, 75-82.

15. Nishikawa, A.H. and Bailon, P. (1975) Anal. Biochem. 68,274-280.

16. Hofstee, B.H.J. and Otillio, N.F. (1978) J. Chromatogr. 161, 153-163.

17. Klotz, I.M. (1970) Arch. Biochem. Biophys. 138, 704-706

18. Hofstee, B.H.J. (1975) Biochem. Biophys. Res. Comm. 63, 618-624.

19. Hofstee, B.H.J. (1979) Biochem. Biophys. Res. Comm. 91, 312-318.

20. Putnam, F.W. (1977) in The Plasma Proteins Putnam, F.W.ed. 2nd ed. Vol. 3, p. 128, Academic Press, New York.

21. Slifkin, M.A. (1971), Charge Transfer Interactions of Biomole-cules, Academic Press, London.

22. Porath, J., Carlsson, J., Olsson, I. and Belfrage, G. (1975) Nature (London) 258, 598-599.

23. Proceedings 3rd Intern. Symp. on Affinity Chromat. and Mol. Interact. (1979) (Strasbourg, June 1979, J.M. Egly, ed.) INSERM, Paris.

24. Hofstee, B.H.J. and Otillio, N.F. (1978) J. Chromatogr. 159 57-69.

25. Axen, R. and Vretblad, P. (1971) Acta Chem. Scand. 25, 2711-2716.

26. McLaren, A.D. (1954) J. Phys. Chem. 58, 129-137.

27. Hofstee, B.H.J. (1962) Biochim. Biophys. Acta 55, 440-454.

28. Peterson, E.A. (1970) in Laboratory Techniques in Biochemistry and Molecular Biology (Work, R.S. and Work, E., eds. Vol. 2, pp. 233-378, North-Holland Publishing Comp., New York.

29. Hofstee, B.H.J. (1973) Bioch. Biophys. Res. Comm. 54, 1137-1144.

30. Dahlgren Caldwell, K., Axen, R. and Porath, J. (1975) Biotechnol Bioeng. 17, 613-616; ibid. (1976) 18, 433-438.

31. Hofstee, B.H.J. (1979) in Proc. 12th FEBS Meeting Vol. 52, (E. Hofmann et.al., eds.) pp. 449-481, Pergamon Press, Oxford.

32. Hofstee, B.H.J. (1975) Prepar. Biochem. 5,7-19.

33. Porath, J. (1954) Arkiv Kemi 7, 535-537.

34. Messing, R.A. (1976), Methods in Enzymology 44, 148-169.

35. Hofstee, B.H.J. (1979) Pure and Appl. Chem. (IUPAC), 51, 1537-1548.

USE OF SYNTHETIC CHEMICAL LIGANDS

FOR AFFINITY CHROMATOGRAPHY OF PROTEINS

Scott P. Fulton and Eleanor R. Carlson

Amicon Corporation
25 Hartwell Avenue
Lexington, MA. 02173

Since its introduction in the late 1960's, the technique of affinity chromatography has proven to be an enormously successful and versatile method of purifying proteins for research purposes.[1] The often extremely high specificity of interaction between the column-immobilized substrates, cofactors, inhibitors, effectors, antigens, etc., and their corresponding enzymes, antibodies, or other binding proteins makes affinity chromatography a valuable addition to, or even replacement for, conventional separation methods, such as salt precipitation, gel filtration, or ion exchange. The widespread use of affinity methods has made purifications to complete homogeneity a routine matter for most proteins, often while maintaining high yield of activity.

In spite of its many demonstrated advantages on the laboratory scale, affinity chromatography has yet to be used to any appreciable extent in large-scale protein purifications. The reasons for this are primarily related to cost. Affinity media typically cost 10-100 times as much as conventional media to purchase initially because the ligand itself must first be isolated and extensively purified, then somehow covalently coupled to the gel, a process which can involve complex and difficult organic synthesis. Protein binding capacity is often low, making it necessary to use larger columns. Affinity columns are usually quite specialized in their applications, whereas an ion exchange or gel filtration column can be used for a wide variety of different purifications. Finally, the biomolecules used as immobilized ligands in affinity media are often subject to enzymatic or chemical degradation, which limits the ability to regenerate and reuse an affinity column, further adding to the cost.

One approach to dealing with these problems is to replace the biomolecular, "natural" ligands with synthetic chemical ligands. Two basic classes of such synthetic ligands have been developed for use in purifying proteins. In the first class, the synthetic ligand serves as a mimic of one or more natural ligands, interacting in a similar fashion with the binding sites of the proteins. In the second class, the synthetic ligand is used to form reversible co-valent bonds with a specific chemical structure on the protein, with separation on the basis of the presence and accessibility of the target structure on the protein surface. Examples of both classes of ligands and their applications are discussed in the following sections.

CLASS ONE SYNTHETIC LIGANDS - TRIAZINE DYES

The first synthetic chemical ligand to gain widespread use in affinity chromatography was the dye Cibacron® Blue F3GA (Ciba-Geigy, hereafter referred to as blue A). This material was originally developed as a color-fast textile dye and has been used extensively in the manufacture of blue jeans. It was introduced into biochemical purification methodology as a chromaphore coupled to very high molecular weight dextran for use as a void volume marker in gel filtration. In 1968, while doing gel filtration characterization of yeast pyruvate kinase, Haeckel et al.[2] attempted to determine the void fraction of the column simulta-neously with the K_{av} of the pyruvate kinase by mixing the protein with blue dextran void marker and applying the mixture to a Sephadex® G-200 (Pharmacia) column. Surprisingly, the pyruvate kinase emerged in the void fraction with the blue dextran. Subse-quent investigation showed that the enzyme was binding to the blue A chromaphore and not the dextran and that the two could be separated by gel filtration in a high salt concentration, giving a 3X purification of the enzyme.

Over the next several years a number of other enzymes were purified using gel filtration with blue dextran. In 1973, Ryan and Vestling[3] published a method of covalently coupling blue dex-tran to agarose to form a true affinity medium and used this material to purify various lactate dehydrogenases. In 1974 Heyns and DeMoor[4] were able to couple the blue A dye directly to agarose, without the dextran bridge. They used this medium to purify rat erythrocyte 3(17)-β-hydroxy-steroid dehydrogenase 220-fold with 50% yield in a single step.

The use of the blue A dye-ligand for protein purification then began to grow rapidly. To date, there have been well over 200 published purifications using this ligand, with novel uses appearing constantly.[5,6] Many of these applications (>80) have involved dehydrogenase and kinase enzymes, but a wide range of other proteins have been purified, including nucleic acid binding enzymes (>15), CoA enzymes (>7), phosphatases, transferases, synthetases, isomerases, sulfhydrolases, serum proteins (α-feto-protein, coagulation and complement factors, steroid carrier proteins, albumin), receptors, myosin, and interferons (human and other mammal leukocyte, fibroblast and lymphoblastoid).

The tremendous success with the blue A dye-ligand prompted a number of investigators to screen among the hundreds of other reactive textile dyes for ligands with superior properties to the blue A for particular applications. Baird et al.[7] found Procion® Red HE-3B (ICI, hereafter referred to as red A) to be outstanding for carboxypeptidase G purification. Watson et al.[8] found that red A was more selective than blue A for NADP versus NAD enzymes. Later work has shown that this dye-ligand is especially effective for alkaline phosphatase, dihydrofolate reductase, 5,10-methylene-tetrahydrofolate reductase, hexokinase, some tRNA synthetases, and plasminogen.

Recently the Amicon Corporation, after extensive dye-ligand screening, introduced three new dye-ligands with different and use-ful properties. Orange A is unusual in the large number of enzymes and other proteins that do not bind, which contrasts sharply with the other dye-ligands. Proteins which do bind, such as carnitine acetyltransferase, chondronectin, citrate synthetase, some comple-ment factors, lactate dehydrogenase, and phosphoglycerate kinase, show outstanding purification factors. The green A dye-ligand is the opposite--more proteins typically bind, more tightly, than to the blue A. This gel is particularly useful for removal of con-taminants from non-binding proteins in "negative chromatography." The blue B dye-ligand has the relatively bulky copper phthalocyanine chromophore and a much lower maximum concentration coupled to the gel. This gives it somewhat weaker binding properties in general, which can permit good recovery of proteins which bind too tightly to the other gels, such as pregnancy-specific β-globulin.

The partial chemical structures of the blue A, red A, and orange A dye-ligands are shown in Figures 1-3. All of the dye-ligands used to date consist of aromatic hydrocarbons substituted heavily with sulfonates, amines, hydroxyls, and other polar groups, joined together by secondary amines or diazo bonds, with coupling to the gel matrix most commonly through an ether linkage to a tri-azine ring, originally substituted with one or two active chlorines.[9]

Considerable work has gone into studying the mechanism of action of the triazinyl dye-ligands. Electrophoresis, enzyme inhibition, bind equilibrium, absorbance and circular dichroism spectral, NMR, thermal and even X-ray crystallographic studies with a wide variety of proteins[5] have shown that although there is con-siderable variation with different dye-protein systems, in many

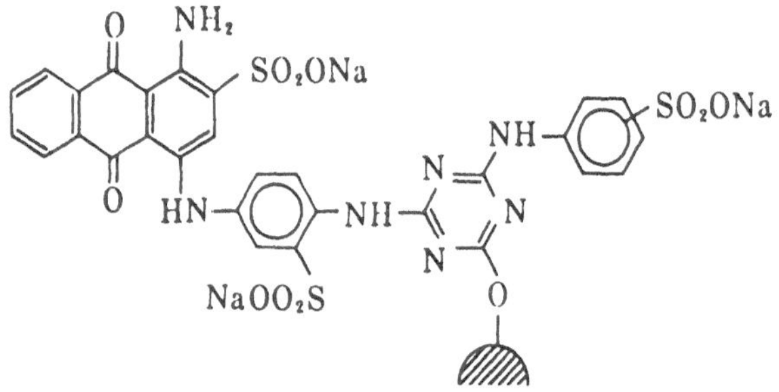

FIGURE 1:   Blue A dye-ligand

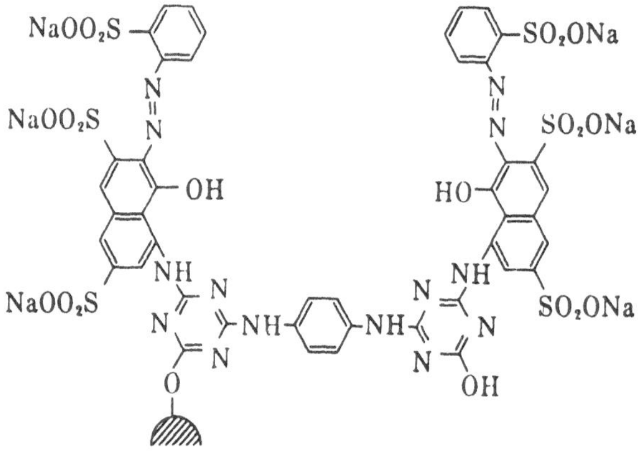

FIGURE 2:   Red A dye-ligand

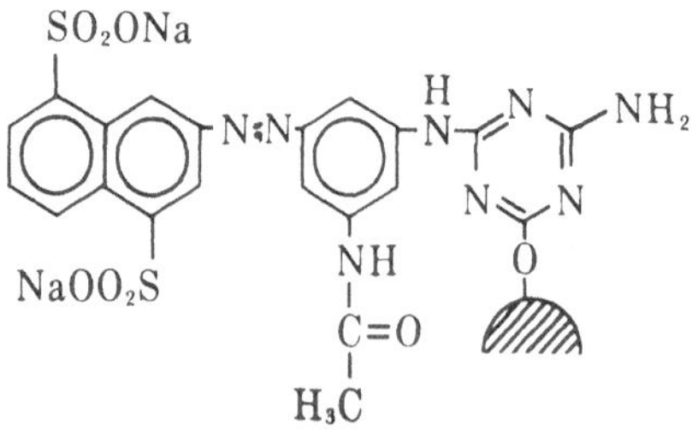

FIGURE 3:   Orange A dye-ligand

cases the dye-ligand binds to the substrate or cofactor binding
site in a manner analogous to the natural ligand. Figure 4 shows
the remarkable similarity between a typical nucleotide cofactor
(NAD[+]) and blue A. In particular, the sulfonate groups on the
dye-ligand occupy similar positions to the phosphate groups on
the NAD[+], and there is considerable similarity in the alignment
of the π-electron planes. The structural flexibility of the dye-
ligands clearly contributes the ability to bind in a wide variety
of protein binding sites for small, anionic biomolecules.

    Early in the development of the blue A dye-ligand, Thompson
et al.[10] proposed that the molecule was specific for proteins containing the so-called "dinucleotide fold," a nucleotide-binding
domain found in a large number of dehydrogenases, kinases, and
other enzymes. Certainly most or all of the proteins known to have
this structure have been found to bind to the blue A. However, so
many other proteins bind that do not have the dinucleotide fold,
such as serum albumin, that the usefulness of the blue A as a
specific probe for this structural feature has been called into
question.[11] The binding interaction of the dye-ligands and proteins
can be very specific (sometimes even very small structural modifications of the dye can prevent binding), but the degree of specificity appears to be on an individual protein-by-protein, dye-by-dye
basis. The use of these ligands, therefore, has been, and probably
will continue to be, on a highly empirical basis.

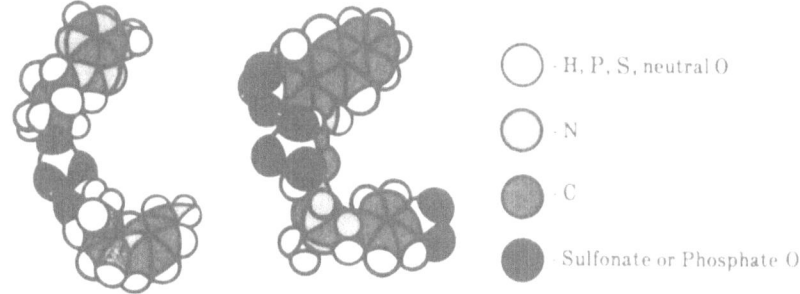

H, P, S, neutral O

N

C

Sulfonate or Phosphate O

FIGURE 4: CPK space-filling models of NAD[+] (left) and blue A
          dye-ligand (right) after [10].

In use, the dye-ligand media display great versatility and
adaptability. A simple screening technique is used to determine
the best dye-ligand for the particular protein being purified in
the particular extract. The purification may then be optimized by
careful selection of binding conditions and elution technique to
give significant improvements in yield and purification factor.

Significant binding conditions include pH, ionic strength,
protein load and concentration, and the presence or absence of
divalent cations, such as magnesium or calcium. Binding to the
ligand can occur over a wide range of these parameters, and protein
stability can be a more important consideration than the binding
characteristics. Most buffer salts and additives, such as mercaptoethanol or glycerol, used to improve recovery of activity, have
little or no effect on the binding to the dye-ligand.

The protein binding capacity of the dye-ligand gels is in the
range of 5-20 mg protein/ml gel, depending upon the protein and
the extract in which it is contained. Often competitive binding
effects are important so that prior purification steps have a
critical influence. For reasons not entirely understood, dye-
ligand gels often have binding capacities 5-100 times those of
equivalent natural ligand media.

Elution can be effected by increasing the ionic strength, or
introducing low (1-20 mM) concentrations of specific eluents
(substrates, cofactors, inhibitors, etc.), or using chaotropic
agents, such as thiocyanate or ethylene glycol. Specific eluents
usually give the highest purification factor, but finding the
right eluent or combination of eluents can be difficult. Ionic
strength elution (0-2 M is usually sufficient) is often more cost-
effective and, when performed with a gradient, can give excellent
results. Chaotropic eluents are necessary when hydrophobic interactions between the dye-ligand and protein predominate and other
elution methods fail.

Other useful techniques using dye-ligands include rechromato-
graphy, with different elution methods, and "negative chromatography," in which the protein of interest does not bind but the
contaminants do. Tandem negative-positive column systems have also
been very useful in a number of cases.

The dye-ligand gels are extremely chemically resistant and can
be effectively regenerated in situ with a 6-8 M urea/0.5 M NaOH
solution. Lifetimes of 30 or more runs are not uncommon and, with
proper care, a column can last well over a year. If extremely
impure samples are used, containing a high content of lipids or
lipoproteins, the capacity and binding strength may drop off
sharply after only one run. This may usually be prevented by
performing a preliminary ammonium sulfate fractionation or

solvent extraction, or the column itself may be restored by exchange into a chloroform/methanol solution.

An example of the use of dye-ligand chromatography to solve
a typical problem is the separation of rabbit muscle pyruvate
kinase (PK, EC. 2.7.1.40) from lactate dehydrogenase (LDH, EC.
1.1.1.27), and the simultaneous purification of both enzymes.
Both enzymes are of commercial interest and are often used together
in a coupled assay system. However, because of the coupled reactions, it is critical to produce each enzyme substantially free of
contamination by the other.

Table 1 shows the results of a screening study for the enzymes
with five dye-ligand agarose media (Matrex^{TM} Gels, Amicon). Each
2-ml column was equilibrated with the starting buffer (20 mM Tris/
HCl, pH 7.5). A 0.5-ml aliquot containing the amounts of protein
and enzymes shown was applied to each column. The columns were
then washed with 10 ml of starting buffer, followed by 10 ml of
elution buffer (1.5 M KCl in starting buffer). The sample wash and
elution fractions for each column were assayed for total protein,
PK, and LDH. Enzymes recovered in each fraction are expressed as
percent of the sample applied. The control column contains gel
with no dye-ligand.

PK bound completely to the blue A, red A, and green A columns,
partially to the blue B column, and very little to the orange A.
Elution of PK was complete only with the blue A and red A gels.
The apparent increase in PK activity in the elution fraction,
particularly seen with the red A, was repeatable and may be due to
removal of an inhibitor, a competing enzyme, or a critical protease.

Table 1. Dye-ligand Screening Results

		Control	Blue A	Red A	Orange A	Green A	Blue B
Wash	PK	93%	0%	0%	86%	0%	50%
	LDH	100%	0%	0%	1%	0%	6%
Elution	PK	0%	122%	203%	10%	48%	64%
	LDH	1%	70%	102%	88%	32%	109%
Purification	PK	--	3X	3X	--	1.5X	2.5%
	LDH	--	1.5X	1.5X	3X	1X	4X

Samples - 2.5 mg protein, 7.9 units PK, 16.3 units LDH

The LDH bound well to all of the columns except blue B and could be eluted with full recovery from all except the green A. The orange A and blue B gels showed the highest purification factors for LDH.

The screening results suggested an interesting approach to solving the purification problem. The orange A column can be used initially to bind the LDH, leaving the PK in the initial wash peak. The orange A wash can be applied directly to a red A column, which will bind the PK. The orange A and red A columns can be eluted separately by salt gradients, producing purified LDH and PK, respectively.

Appropriate column loadings were determined by frontal analysis[1,5] to be approximately 15 mg crude protein/ml gel. Crude extract was applied to each of the columns, and a 0-1 M KCl gradient was used to elute the PK and LDH. As seen in Figure 5, the two enzymes were separated by the gradient on each of the gels. This is fortuitous since in the tandem separation it will give a double separation of the two enzymes, which should effectively eliminate a cross-contamination.

Table 2 shows the results of a pilot purification on the 2-ml orange A and red A columns. Initially, the two columns were coupled together, with the eluate of the orange A feeding into the top of the red A. After washing the sample through the orange A gel with buffer, the columns were separated and the red A given an additional wash. The columns were then eluted on separate salt gradients. Final yield of each enzyme was high, each was purified to a considerable extent, and each was substantially free of con- tamination by the other.

Table 2. Final PK and LDH Purification

Starting Material: 27 mg protein
                     63 units PK (2.3 U/mg)
                     52 units LDH (1.9 U/mg)

Orange A Eluate: 42 units LDH (81% yield)
                     30 U/mg (16X purification)
                     PK activity < 0.3% of LDH

Red A Eluate: 50 units PK (80% yield)
                     10 U/mg (4.5X purification)
                     LDH activity < 0.5% of PK

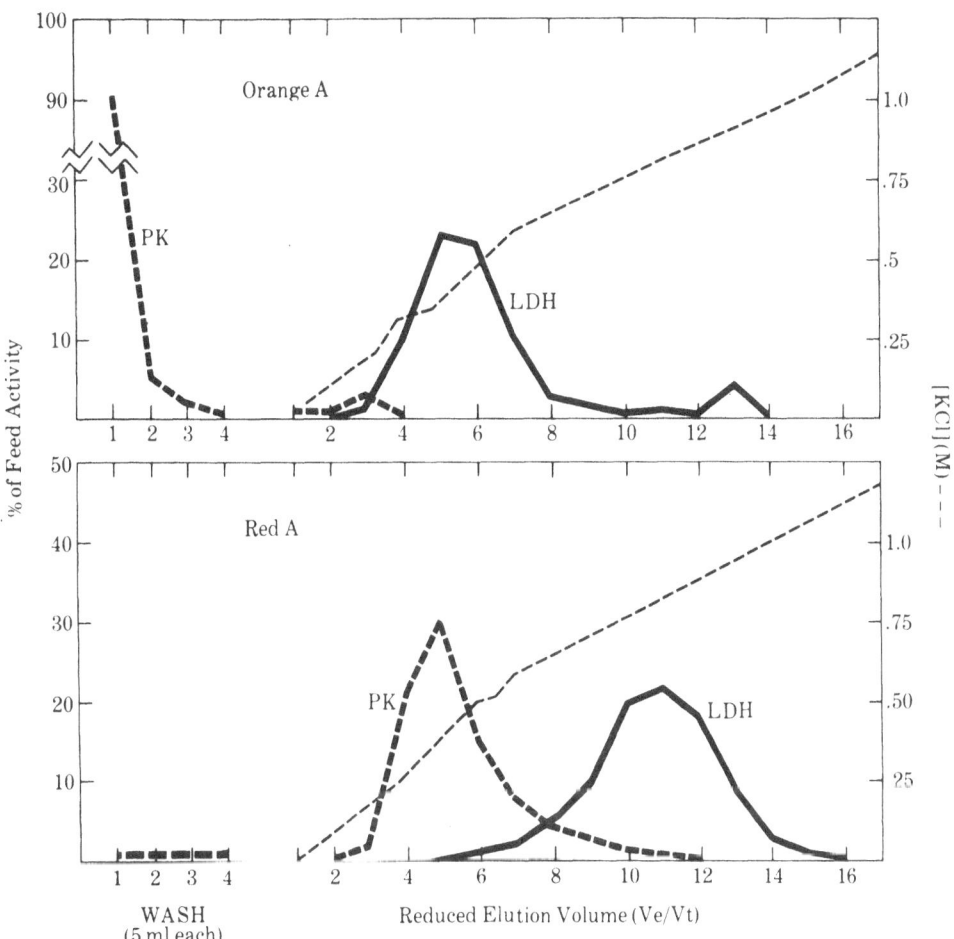

FIGURE 5: Salt gradient elution of crude rabbit muscle acetone powder (Sigma) from Mātrex Gels Orange A and Red A. Samples (20 mg protein, 1 ml) were applied to each 2-ml column, which was washed in buffer (20 mM Tris/ HCR, pH 7.5), followed by a 0-1 M linear KCl gradient.

CLASS TWO SYNTHETIC LIGANDS - PHENYL BORONATE

The second type of synthetic affinity ligands are those which can form a reversible covalent bond with a particular chemical structure on the surface of the protein being purified. The first such ligands to appear were directed at the sulfhydryl groups of cysteines. Chloromercurobenzoate,[12] glutathione, and thiopropanol[13] ligands have all been used successfully for covalent chromatography of thiol-containing proteins and peptides. Binding occurs in the normal biological pH range,[4-8] and elution is generally effected by using ~10 mM concentrations of competing thiol compounds, such as cysteine, mercaptoethanol, or dithiothreitol.

Current work at Amicon has focused on another covalent chromatography ligand--phenyl boronate. It has been known for nearly a century[14] that boric acid is able to form complexes with various polyols, and this fact has been used for some time in the study of carbohydrate structure.[15] In 1931 Seaman and Johnson[16] synthesized m-aminophenyl boronate for study as a potential bacteriostat, but it was not until 1970 that Weith et al.[17] immobilized this compound (on cellulose) and used it as an affinity ligand for sugars and nucleosides. Others have used this ligand, coupled to various supports, for the purification of nucleic acids (especially tRNA)[18] and catechols.[20] Only very recently has the phenyl boronate ligand been used for covalent chromatography of glycoproteins.

Two types of interaction can occur between boronate ligands and molecules of biochemical interest. Figure 6 shows the basic interaction with polyols. At a pH between 8 and 9, the boronate is hydroxylated to convert from an uncharged, planar, trigonal form to an anionic, tetrahedral form. The tetrahedral form is able to reversibly exchange hydroxyls as shown with any molecule containing a 1,2-diol which can assume a coplanar (cis) arrangement. Thus, any such molecule, from a catechol to a membrane glycoprotein, can be retarded or bound to a phenyl boronate column.

A second type of interaction (Figure 7) can occur with amines. The highly electrophilic trigonal boronate can form a charge-transfer complex with the free electron pair of the amine. Note that this type of complex has no net charge. Mixed interactions with hydroxyl-containing amines, such as Tris buffer or glucos-amine, are also possible and, indeed appear to be quite strong.

Requirements in operating conditions for the phenyl boronate ligand are somewhat more stringent than for the dye-ligands. The pH must be above 8.0 for effective binding to occur. Certain buffers, such as Tris, ethanolamines, and borate, cannot be used because they interfere with binding. Use of low (0-10°C) temperatures will usually enhance binding. Divalent cations are

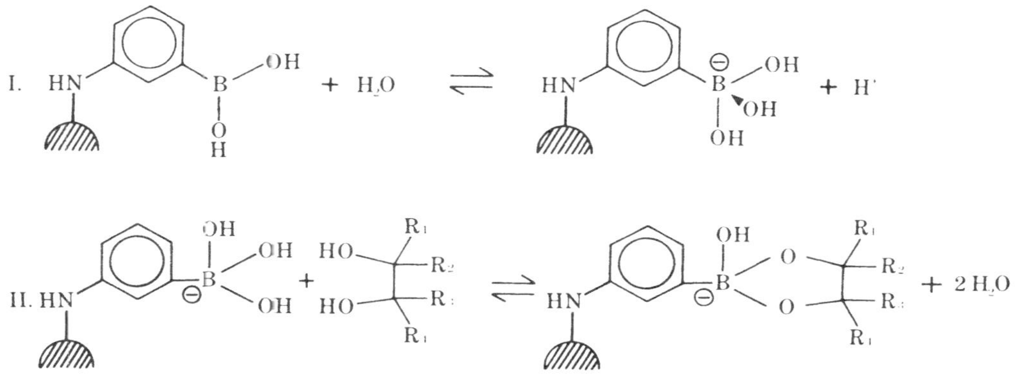

FIGURE 6:   Interactions of phenyl boronate ligand with coplanar 1,2-diol compounds.   I. Conversion of boronate from trigonal to tetrahedral form.   II. Reversible hydroxyl exchange with diol boronate.

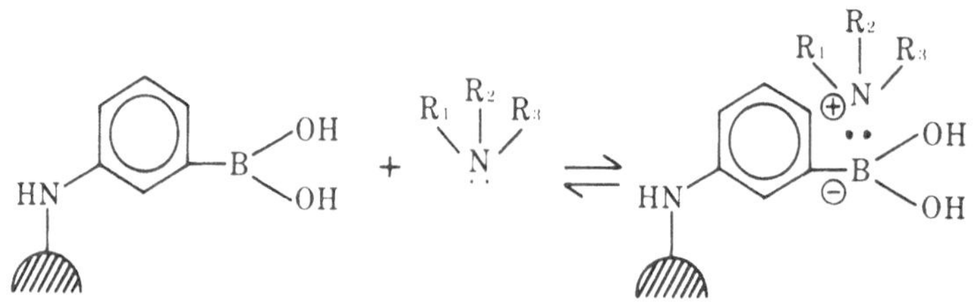

FIGURE 7:   Charge-transfer interaction of boronate with amines.

essential for binding in some cases. Other conditions, however,
such as ionic strength and most buffer additives, have little or
no effect on binding.

   A critical consideration with the phenyl boronate ligand is
the possibility of "nonspecific" hydrophibic and ionic interactions,
which add to or subtract from the "specific" boronate interaction.
If care is taken to prevent the binding of non-diol-containing
molecules, these interactions can be very useful in that they
permit separations of different diol-containing molecules which do
bind.

   Figure 8 shows the effect of pH on the binding of a crude
horseradish peroxidase mixture on both a phenyl boronate and an
equivalent phenyl (no boron) gel. Binding of the basic isoenzymes
($\sim$80% of OD_{403}) occurs at all pH's to the phenyl boronate gel, but
only at low pH's to the phenyl gel. Elution with 15 mM sorbitol
is effective only on the phenyl boronate gel at pH's above 8.0.
This illustrates the effect of hydrophobic binding. The acidic
isoenzymes ($\sim$20% of OD_{403}) do not bind in this experiment (where
no Mg^{2+} is present) but will bind in 10 mM $MgCl_2$ and can be
selectively eluted with 10 mM EDTA. This is presumably due to a
repulsive effect between acidic groups close to the binding diols
and the anionic boronate, which can be bridged by the divalent
magnesium ions.

   Elution can be performed in a variety of ways. Sometimes the
binding is weak enough to allow elution by washing the column with
the starting buffer. If long columns are used, binding molecules
are separated on the basis of different retardation. Shifting to
a lower pH will elute some molecules, although with many glycoproteins this will induce hydrophobic binding to the phenyl
component of the ligand. Molecules which require Mg^{2+} or Ca^{2+} for
binding can be eluted with EDTA. Competing polyols or hydroxyl

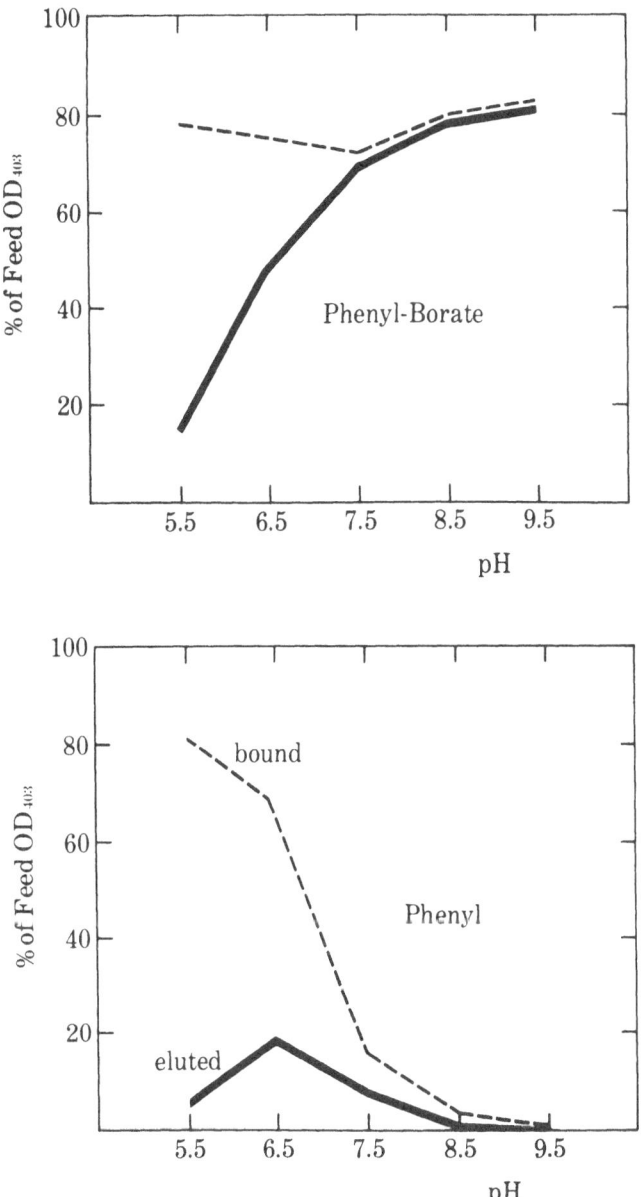

FIGURE 8:   Binding and elution of partially purified horseradish
            peroxidase (6 mg protein) on 2-ml columns of Matrex Gel
            Phenyl Boronate (top) and an identical gel with no
            boronate present (bottom).

amines (5-100 mM), such as mannitol or Tris, can be used in step or
gradient elutions. Urea or ethylene glycol at high concentrations
can be used where the combined boronate and hydrophobic interactions
prevent elution by other means.

Extensive reuse of the phenyl boronate gels is possible if the
gel is regenerated with the urea/NaOH solution used for the dye-
ligands. It should be noted that many of the coupling methods used
form an amide linkage to the amino phenyl group, which is hydrolyzed
in strong base. Proper coupling chemistry alleviates this problem
and allows the same lifetimes observed with dye-ligand gels.

An example of the use of phenyl boronate media is the purification of glucose oxidase (EC. 1.1.3.4) from A. niger. The major
problem in the production of this commercially important enzyme is
the removal of catalase (EC. 1.11.1.6), which has such a high turn-
over number that even trace amounts of protein have a very high
enzymatic activity.

Initial screening experiments showed that both the catalase
and the glucose oxidase bind to phenyl boronate agarose (Mātrex
Gel Phenyl Boronate, Amicon), but that the glucose oxidase requires
10 mM magnesium. The glucose oxidase can be eluted separately from
the catalase by 10 mM EDTA.

Figure 9 shows a larger scale purification using this method.
The glucose oxidase was recovered with 48% yield and a 3X purification. More importantly, the catalase activity was reduced from
2500% of the glucose oxidase activity (in equivalent units) to less
than 9%. A second pass through a phenyl boronate column equilibrated
in the elution buffer can be used negatively to strip most of the
residual catalase without reducing the glucose oxidase yield.

The direct covalent chromatography of glycoproteins on phenyl
boronate media is a relatively new idea, and a few purifications
have yet appeared in the literature. Exceptions include nucleo-
peptides,[20] and pancreatic lipase.[21] Unpublished purifications
include glycosylated hemoglobins,[22] glycopeptides, and horseradish
peroxidase. With the commercial availability of inexpensive phenyl
boronate media designed for high molecular weight samples, use of
this novel "synthetic lectin" should increase.

The phenyl boronate ligand has another use which, while not
strictly a synthetic ligand application, enables one to overcome
many of the problems encountered with large-scale natural ligand
affinity chromatography. As shown in Figure 10, the phenyl
boronate is used to temporarily immobilize a natural ligand
possessing the required diol structure, which, in turn, binds the
protein of interest. In general, the natural ligand is applied to
the column at a certain critical concentration, the sample is

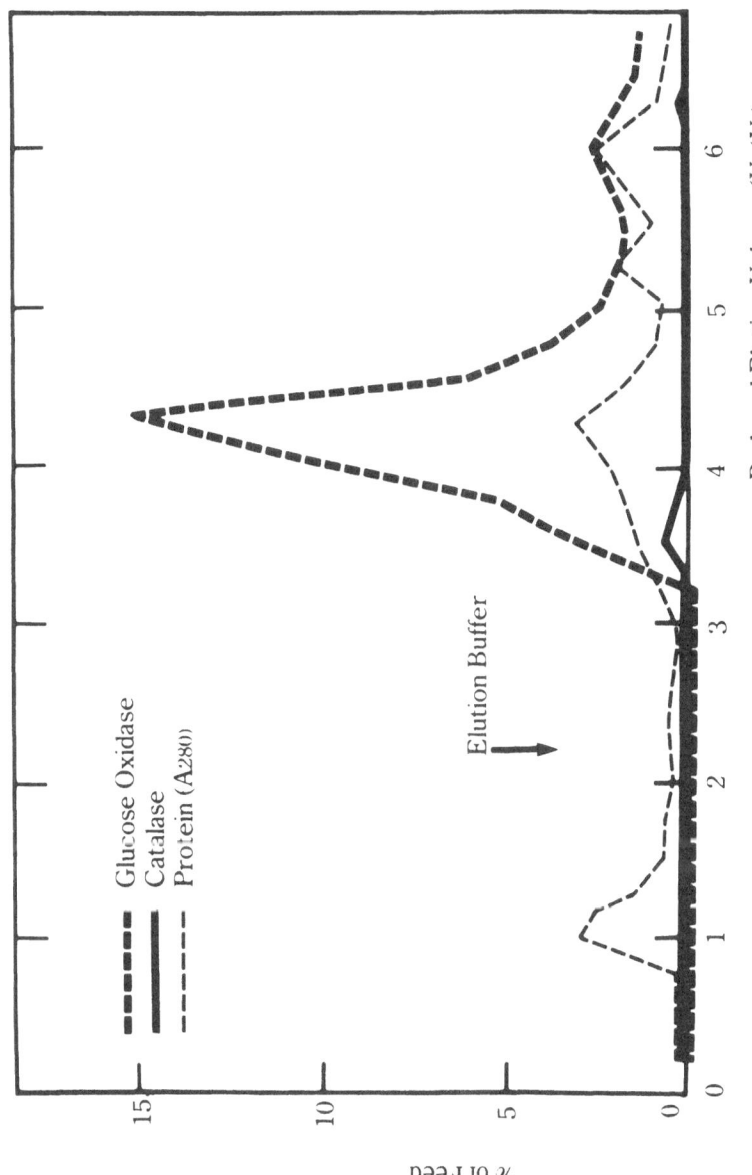

FIGURE 9: Purification of glucose oxidase from A. niger on a Mātrex Gel Phenyl Boronate column (1 x 25) cm, 20 ml. Starting Buffer- 50 mM HEPES pH 8.5, 10 mM MgCl$_2$. Elution Buffer- 50 mM HEPES pH 8.5, 10mM EDTA.

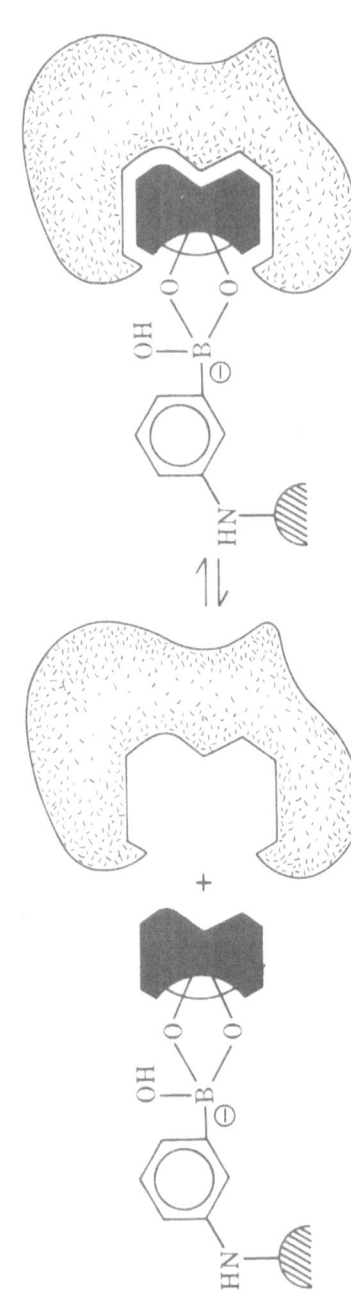

FIGURE 10:   Schematic of indirect or "piggyback" chromatography using phenyl boronate ligand gel.

applied, the column is washed with a buffer containing the ligand, followed by a buffer without the ligand, which elutes the protein.

This method has the advantages that no expensive synthesis of specific affinity media is required for each application and that the ligand may be renewed for each run. This second point is especially important when the protein of interest or contaminating proteins are able to degrade the ligand. With conventional affinity media, this would necessitate packing a new column for each run, whereas the phenyl boronate column can be simply regenerated and recharged with ligand. To date, this method has been utilized to purify various dehydrogenases,[23] UDP-glucose pyrophosphorylase,[24] and α-glucosidase.

Synthetic ligand affinity chromatography is a versatile, efficient, and cost-effective means of purifying proteins. To the purist, perhaps, the method has the drawback that the interactions involved are not strictly "biospecific" in the manner of conventional affinity techniques. It is true that this limits the analytical applications of the method. Synthetic ligand media, however, can be as effective as natural ligand media in providing high purification factors and high yield. Their low-cost, high binding capacity, operational versatility, and durability make them the only economically viable means of bringing the high selectivity of affinity chromatography to large-scale industrial purifications of enzymes and other proteins.

REFERENCES

1. Lowe, C. and Dean, P., "Affinity Chromatography." John Wiley and Sons, New York (1974).

2. Haeckel, R., Hess, B., Lauterborn, W., and Wuster, K. Hoppe-Seyler's Z. Physiol. Chem. 349:699 (1968).

3. Ryan, L. and Vestling, C. Arch. Biochem. & Biophys. 160:279 (1974).

4. Heyns, W. and DeMoor, P. Biochim. Biophys. Acta. 358:1 (1974).

5. Fulton, S. "Dye-ligand Chromatography." Amicon Corporation, Lexington, MA. (1980).

6. Dean, P. and Watson, D. J. Chromatog. 165:301 (1979)

7. Baird, J., Sherwood, R., Carr, R. and Atkinson, A. F.E.B.S. Letters. 70:61 (1976)

8. Watson, D., Harvey, M. and Dean, P. Biochem. J. 173:591.

9.   Beech, W., "Fibre Reactive Dyes," SAF International, New York (1978).

10. Thompson, S., Cass, K., and Stellwagen, E. Proc. Nat. Acad. Sci. USA. 72:669 (1975).

11. Stellwagen, E., Affinity chromatography using immobilized anionic dyes, in "Affinity Chromatography and Molecular Interactions," J.-M. Egly, ed., INSERM, Paris (1979).

12. McDonagh, J., Waggoner, W., Hamilton, E., Hindernack, B., and McDonagh, R. Biochim. Biophys. Acta. 446:345 (1976).

13. Axen, R., Drevin, J., Carlsson, J. Acta Chem. Scand. B29: 471 (1975).

14. Magnanini, G. Gazz. Chim. Ital. 201:428 (1890).

15. Boesekan, J. Advan. Carbohyd. Chem. 4:189 (1949).

16. Seaman, W. and Johnson, J. J. Amer. Chem. Soc. 53:711 (1931).

17. Weith, H., Wiebers, J. and Gilham, P. Biochemistry. 9:4396 (1970).

18. Schott, H., Rudloff, E., Schmidt, P., Roychoudhury, R. and Kossel, H. Biochemistry. 12:932 (1973).

19. Higa, S., Suzuki, T., Hoyoshi, A., Tsuge, I. and Yamaminura, Y. Anal. Biochem. 77:18 (1977).

20. Annamali, A., Pal, P. and Colman, R. Anal. Biochem. 99:85 (1979).

21. Akparov, V., Nutsubidze, N., Rotanova, T. Bioorg. Khim. 6:609 (1980).

22. Dean, P., Brown, P. and Bouriotis, V. unpublished results.

23. Dean, P., Qadri, F., Jessup, W., Bouriotis, V., Angal, S., Potuzak, H., Letherbarrow, R., Miron, T., George, E., and Morgan, M. Design faults in affinity chromatography, in "Affinity Chromatography and Molecular Interactions," J.-M. Egly, ed., INSERM, Paris (1979).

24. Maestas, R., Prieto, J., Kuehn, G. and Hagemen, J. J. Chromatog. 189:225 (1980).

SEPARATION OF AMINES AND ACIDS BY

STERICALLY HINDERED POLYMERS

Y. Okamoto, M. Khojasteh, and C. J. Hou

Department of Chemistry and Polymer Research
Institute, Polytechnic Institute of N.Y.
333 Jay Street
Brooklyn, NY 11201

INTRODUCTION

In recent years a number of polymeric reagents
have been extensively utilized as specific polymeric
adsorbents for biologically active compounds and
metallic ions[1,2]. In general, for a selective separation of substrates, specific polymeric reagents which
are capable of selectively binding one or a few species
in a complex mixture are chosen. This step would
be followed by isolation of the polymer bound compound
from the mixture.

The chemistry of highly hindered 2,6-di-t-butyl
phenol has been the subject of a number of investigations in past years[3,4]. The noteworthy properties of
the compound include (a) the stability of the phenoxy
radicals, (b) a failure to exhibit normal phenol
properties. A highly hindered pyridine, 2,6-di-t-
butyl pyridine also possesses interesting and unusual
properties. The compound was first synthesized by
Brown and Kanner[5]. They showed that the compound did
react with protonic acids but did not react with Lewis
acids such as BF_3 or with CH_3I by conventional
procedures. Thus, we have initiated a program to
synthesize polymers containing these highly hindered
moieties and to study their applications for the
separation of amines and acids.

EXPERIMENTAL

The copolymers of 2,6-di-t-butyl 4-vinyl phenol[6] (1) and 2,6-di-t-butyl 4-vinylpyridine (2) with styrene and the homopolymer of (1) were prepared by the following schemes.

Synthesis of 1

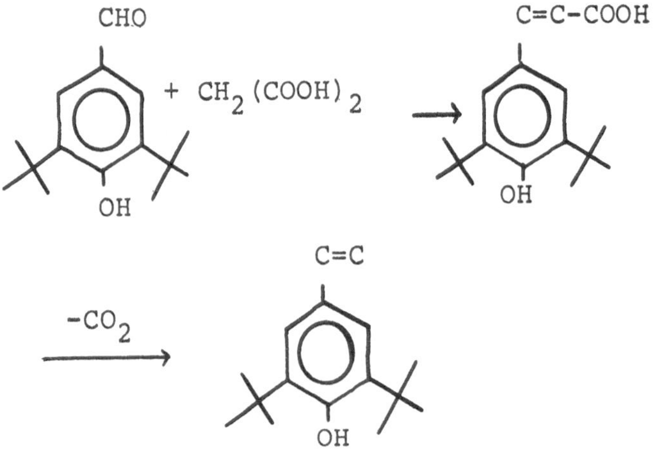

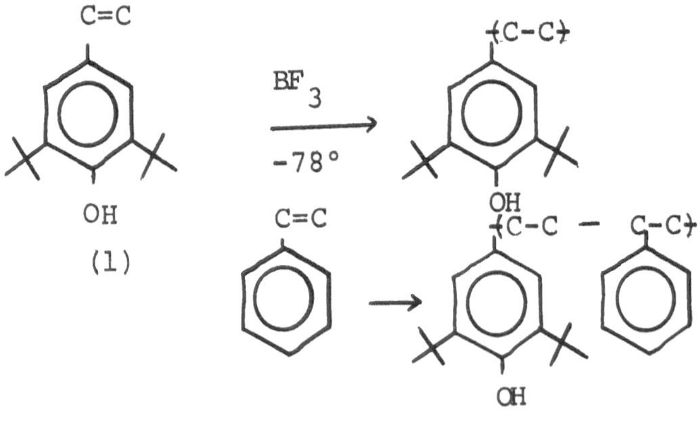

Synthesis of 2

The overall yield from 3,5-di-t-butyl 4-hydroxy-benzaldehyde to (1) was 40-50%. The metalation step of 4-methyl-2,6-di-t-butyl pyridine was found to be very difficult and the yield to (2) was very poor ≈5%. The alternative synthetic procedures for (2) are being studied. The contents of the phenol and pyridine monomers in the copolymers were determined by UV spectrophotometry. The powdered polymers were used in suspension in hydrocarbon solvent. The general pro-cedure for the separation was as follows: A known amount of the powdered polymer was mixed with a hexane solution of amines and the mixture was stirred for 0.5- 1 hr at room temperature and then centrifuged. The solution before and after mixing with the polymer was analyzed by gas chromatography. For determination of the concentration of protonic acid, a tracer technique (using tritium) was used. The polymer used

for the separation was regenerated by treatment of
acidic or basic 95% ethanol.

RESULTS AND DISCUSSION

      In general, the phenol compound reacts with strong
and weak bases through ionic or hydrogen bonding
formation.   There is no particular selectivity among
the reactions of a normal phenol and a series of
primary, secondary and tertiary amines.   But,
sterically hindered phenol 2,6-di-t-butyl 4-ethyl-
phenol interacts with only amine and pyridine compounds
that are not sterically hindered.   Thus, the polymers
containing 2,6-di-t-butyl phenol moiety were applied
in the separation of amine and pyridine compounds. The
typical results are summarized in Tables 1 and 2.

TABLE 1

Separation of Pyridine Derivatives
After Treatment

	Initial mole fraction(%)	Homo- polymer	Copoly- mer[a]	Poly(4- vinyl phenol)[b]
4-Ethyl- pyridine	17	1	3	15
2,6-Di- methyl pyridine[c]	83	99	97	85

a) Copolymer 2.5 g used (phenol moiety
   concentration 0.0046 mole)
b) Obtained from Maruzen Oil Co., Resin-M)
c) In hexane solution (0.001 mole). After treatment
   with the polymers, the pyridine was receovered
   95∿97%.   After treatment with poly(vinyl phenol),
   the pyridine was recovered 5∿10%.

TABLE 2

Separation of Amines

	Initial mole fraction(%)	After treatment		
		Homo-polymer	Copoly-mer[a]	Poly 4-vinyl phenol
n-butyl amine	5	0.1	1	5
tri-n-butyl amine[b]	95	99.9	99	95

a) Copolymer 2.0 g used (phenyl moiety concentration 0.0056 mole)
b) Tri-n-butylamine hexane solution (0.001 mole) used. After treatment with the polymers, the amine was recovered 97∿98%. With poly(4-vinylphenol) the amine recovered only 5∿7%.

    The copolymer of 2,6-di-t-butyl 4-vinyl pyridine with styrene (the pyridine concentration is 25 mole%) was applied to remove any trace amount of HCl present in acid chloride. The powdered polymer was mixed with the hexane solution of acetyl chloride or benzoyl chloride which contained a trace amount of HCl (TCl). The mixture was stirred for 0.5-1 hr at room temperature. It was found that the HCl could be removed quantitatively from the solution and the pure chloride was recovered.

REFERENCES

1) C.C. Leznoff, Chem.Soc.Rev., 3, 65 (1974).
2) M. Kraus and A. Patchornik, Chemtech, Feb. (1979).
3) G.M. Coppinger and T. W. Campbell, J. Am. Chem.Soc., 75, 734 (1953).
4) C.D. Cook and B.E. Norcross, J. Am. Chem. Soc., 78, 3797 (1956).
5) H.C. Brown and B. Kanner, J. Am. Chem. Soc., 88, 986 (1966).
6) D. Braun and B. Naier, Die Makro. Chemie, 167, 119 (1973).
7) A.G. Anderson and P.J. Stang, J. Org. Chem., 41, 3034 (1976).

POLYMERIC SEPARATION MEDIA. NEW FUNCTIONALIZED POLYMERS FOR THE SELECTIVE REMOVAL OF HAPTENS FROM COMPLEX ORGANIC MIXTURES

Jean M. J. Fréchet and Claude Benezra

Department of Chemistry, University of Ottawa, Ottawa
Ont. KIN-9B4, Canada; and Laboratoire de Dermato-Chimie
Clinique Dermatologique, CHR de Strasbourg, Strasbourg,
F-67091, France

INTRODUCTION

Among the numerous properties of functional polymers, one which has been of interest to us recently is their ability to "fish out"[1] selectively one type of functionality from a complex reaction mixture. Our first contribution[2] to this field was the development of a polystyryl boronic acid resin which was capable of binding selectively cis-diols. For example, the separation of cis 1,2-cyclohexanediol from its trans isomer could be carried out efficiently by treatment of the commercially available mixture of isomers with the boronic acid resin using either a batch or a column technique[2]. Boronate formation occurs exclusively with the cis diol which is thus retained by the polymer from which it can be cleaved later by addition of moist solvent or alcohol. This method can also be used with 1,3 or 1,4 diols and is of particular value with carbohydrate derivatives[3-4]. Other noteworthy applications of functional polymers, in which their ability to extract a component from a complex mixture is used for synthetic purposes, include the preparation of a threaded macrocycle[5], the synthesis of an unsymmetrical tetraarylporphyrin[6], and the preparation of a dideoxy-sugar of biological importance such as methyl α-D-paratoside[1]. More recently, the collaboration between our two laboratories has resulted in the development of a new and very effective method for the removal of haptens from natural oils[7,8]. A number of plant extracts, particularly those obtained from plants of the Compositae family contain sesquiterpene α-methylene-γ-butyrolactones. Since these are known to be allergenic substances, and

117

since the plant extracts are used in the perfume, cosmetics and pharmaceutical industries, it is desirable to find a method to effect complete removal of the allergens through a selective process which would leave all the other components of the oils unaffected.

EXPERIMENTAL STRATEGY

Since the mechanism of allergic contact dermatitis to α-methylene-γ-butyrolactones is thought to proceed through the formation of a covalent bond with the nucleophilic skin proteins, it was felt that inactivation of the allergens could be achieved through their reaction with external nucleophilic moieties via a Michael addition. In theory, these nucleophiles could be soluble low molecular weight compounds such as amines. Indeed, our preliminary results indicate that amines react slowly with the unsaturated lactones, however, the use of such low molecular weight additives result in the inactivation but not the removal of the allergenic substances from the natural oils. Furthermore, the presence of unreacted additives might have a serious deleterious effect on the mixture, both through the possibility of further chemical reactions involving other desirable components of the oil, and through changes in physical, chemical, and olfactic properties. In contrast, the use of an insoluble polymer for the treatment of natural oils is particularly attractive since their Michael addition to the lactones will result in both deactivation and extraction of the allergens. The key removal step will amount to a simple separation of phases: the allergenic substances being retained by the solid polymer, and the liquid phase consisting of allergen-free oil. An added benefit of the polymer treatment is that recovery of the polymer-retained substances is possible provided an appropriately designed polymer is used. Cleavage from the polymer will help confirm the structure of the allergenic substances and will also serve to establish that few or preferably no other substances originally present in the oil have been removed by the polymer treatment. In addition, since numerous unsaturated lactones have been shown to be active antitumour agents[9], this procedure could be of use in the isolation of new active substances or in the preparative extraction of these antitumour agents from natural sources.

The focus of our experimental approach was to prepare and test reactive polymers containing nucleophilic amino-groups. Binding to the polymer would result from a Michael addition to the unsaturated lactone, and cleavage would be achieved by permethylation-elimination.

PREPARATION AND TESTING OF AMINATED POLYMERS.

Initial studies with soluble amines having indicated that they were able to add to the α,β-unsaturated lactones, several insoluble polymers containing amine functional groups were prepared, for

example by chemical modification of crosslinked polystyrene[7,8] which
afforded polymers I-III with capacities of up to 5 mequiv./g.

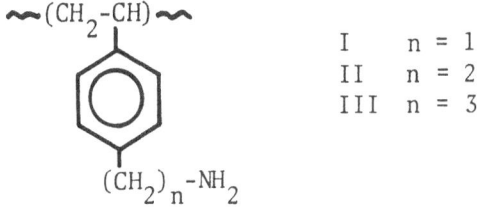

I   n = 1
II   n = 2
III   n = 3

These polymers were prepared both on a macroporous 4% crosslinked
and a microporous 1% crosslinked resin (Amberlite XE-305 and Bio-
beads SX-1, respectively). In general best results were obtained
with the microporous resins which proved to be more reactive. An
alternate approach involved the cyanoethylation of crosslinked
hydroxylated polymers, followed by reduction of the polymeric nitri-
les with lithium aluminium hydride.

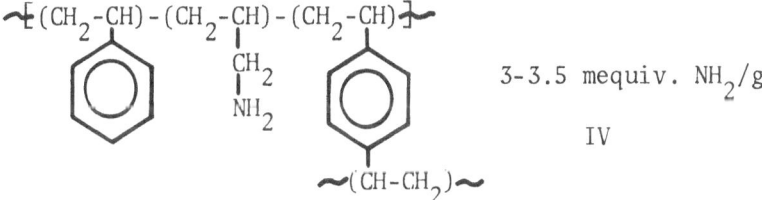

The hydroxylated polymers used were 1% crosslinked polystyrene and
2% crosslinked polyvinyl alcohol.   Perhaps the most inexpensive
and most readily accessible substrate was a resin prepared by ter-
polymerization of acrylonitrile, styrene, and divinylbenzene with
subsequent transformation of the nitrile groups into amines (IV)

3-3.5 mequiv. NH_2/g

IV

The various polymers were tested in the removal of a model unsaturated lactone, isoalantolactone (V), from solutions of known concentrations. The binding was followed by gas chromatography, UV spec-
trometry, and also gravimetrically. The most important experimental
variable was found to be the solvent in which the reaction was
carried out, ethanol giving best results, while the nature of the
polymer did not have any drastic effect on binding.

ISOLATION AND IDENTIFICATION OF COSTUS HAPTENS

     Costus essential oil is obtained by steam distillation of the
dried roots of Saussurea lappa Clarke (a plant of the Compositae

family); it is a light brown viscous liquid with a distinct and
lasting odour. Chromatographic analysis of the oil reveals that it
contains dozens of major components, more detailed studies indicate
that at least two of these compounds are sesquiterpene lactones[10]:
Costunolide (VI), and Dehydrocostuslactone (VII)

Dermatologists became interested in Costus oil when it became apparent that it caused allergic contact dermatitis. The sensitizing
power of Costus oil was attributed to Costunolide and dehydrocostuslactone[11]. Removal of these allergens from the crude oil was
achieved by treatment of a 40-50% solution of the oil in ethanol
with approximately 25% of its weight of an aminoethyl resin (II)[12].
The reaction, followed by gas chromatography, resulted in the
binding of essentially all of the allergens. Typically, polymer
treatment was accompanied by a 13-14% weight loss for the oil.
Testing of the polymer-treated oil on sensitized animals showed that
it had lost its allergenic properties. Cleavage of the lactones
from the polymer was achieved by permethylation of the polymer,
followed by treatment with bicarbonate to effect cleavage. This
reaction was only moderately successful and several successive
treatments (permethylation-base cleavage) were required to recover
80% or more of the bound material[12]. Analysis of the product which
was cleaved from the polymer showed that it was indeed a mixture of
several unsaturated lactones, mainly dehydrocostuslactone (VII) and
rearrangement products of the very labile costunolide (VI). Testing
of the cleavage products on sensitized animals confirmed their
powerful allergenic properties.
Similar results were obtained with another commercial costus extract
which was found to contain approximately 40% of its weight in unsaturated lactones. As was the case with Costus oil, polymer treatment removed all the allergenic unsaturated lactones while leaving
the other components of the extract essentially unaffected.
Examination of the reaction scheme which is shown on the next page
shows one area where improvements are desirable and possible: the
cleavage step is relatively difficult and the polymeric by-product
of the cleavage reaction is a tertiary amine from which (II) cannot
be regenerated in simple fashion. We have attempted to reuse this
polymer by-product by transforming it into the corresponding secondary amine but this reaction, though possible, is of little value.

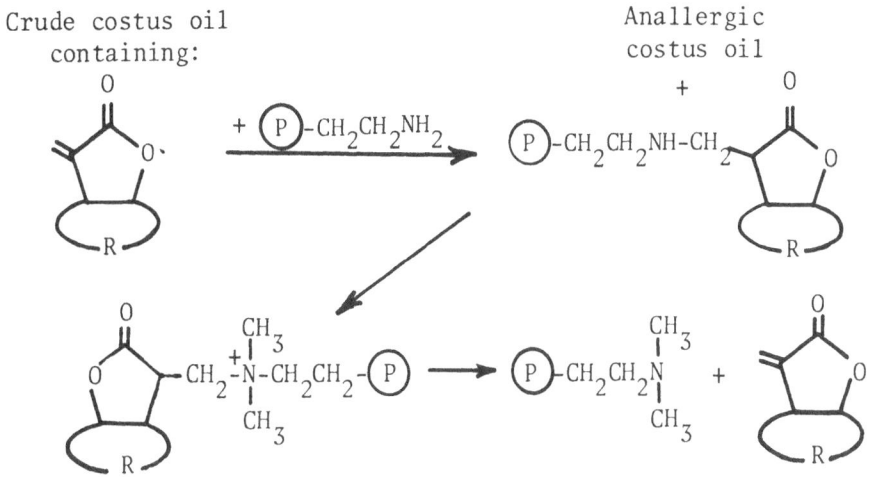

For most applications the benefits derived from the use of a polymer such as II outweigh the inconvenience of having to use a fresh batch of polymer for each treatment.

Although we have not observed significant binding of products other than unsaturated lactones in the treatment of costus oil with poly(aminoethyl styrene), total selectivity cannot be expected with this type of resin in the treatment of mixtures containing other classes of α,β-unsaturated systems. For example, while citral and carvone do not react appreciably with poly(aminoethyl styrene), p-mentha-3,6-diene-2,4-dione and 4,4-dimethyl cyclohexene carboxaldehyde bind rapidly and extensively to the polymer [13].

FUTURE PROSPECTS

The results which are summarized above clearly show the great potential of functional polymers for the separation of specific classes of compounds from complex mixtures. In the case of α-methylene-γ-butyrolactones, aminated polymers can produce excellent results but suffer from their lack of regenerability. Clearly, a fully regenerable polymer would be highly desirable to reduce the unit cost of the polymer treatment trough repeated use. Some pre-liminary studies [13] in this direction are in progress with several systems under consideration. One such polymer which is currently being studied contains carboxylate functionalities on crosslinked polystyrene or methacrylic backbone. If conditions can be found to improve the reactivity of such resins in Michael additions with unsaturated lactones, it is expected that both the polymer and the lactone could be regenerated in one step after use, through a β-elimination as shown on the next page.

First step: binding to the polymer

Second step: cleavage and regeneration

ACKNOWLEDGEMENTS

Thanks are due to our coworkers: A. Cheminat and J.L. Stampf in Strasbourg, and M. de Smet and M.J. Farrall in Ottawa. Financial support by INSERM, NATO, and NSERC is gratefully acknowledged.

REFERENCES

1. J.M.J. Fréchet, Tetrahedron, in press (1980)
2. E. Seymour and J.M.J. Fréchet, Tetrahedron Lett., 3669 (1976)
3. J.M.J. Fréchet, L.J. Nuyens and E. Seymour, J. Am. Chem. Soc., 101, 432 (1979).
    E. Seymour and J.M.J. Fréchet, Tetrahedron Lett., 1149 (1976)
4. K. Krohn, K. Eberlein, and G. Gercken, J. Chromatog., 153, 550 (1978)
5. I.T. Harrison and S. Harrison, J. Am. Chem. Soc., 89, 5723 (1967)
6. C.C. Leznoff and P.I. Svirskaya, Angew. Chem. Int. Ed., 17, 947 (1978)
7. A. Cheminat, C. Benezra, M.J. Farrall and J.M.J. Fréchet, Tetrahedron Lett., 21, 617 (1980)
8. J.M.J. Fréchet, M.J. Farrall, A. Cheminat and C. Benezra, Polym. Prep., 21 (2), 101 (1980).
9. S.M. Kupchan, Fed. Proc., 33, 2288 (1974)
    I.H. Hall, K. Lee, C.O. Starnes, S.A. Eigebaly, T. Ibuka, Y. Wu, T. Kimura, and M. Haruna, J. Pharmaceut. Sci., 67, 1235 (1978)
10. Y.R. Naves, Helv. Chim. Acta, 31, 1172 (1948);
    B. Maurer and A. Grieder, Helv. Chim. Acta, 60, 2177 (1977);
11. J.C. Mitchell and W.L. Epstein, Arch. Dermatol., 110, 871 (1974)
12. A. Cheminat, C. Benezra, M.J. Farrall and J.M.J. Fréchet, submitted for publication.
13. A. Cheminat, C. Benezra, M. de Smet and J.M.J. Fréchet, unpublished results.

COMPARATIVE PERFORMANCE OF MICROPARTICULATE PACKINGS FOR AQUEOUS

STERIC EXCLUSION CHROMATOGRAPHY

Thomas V. Alfredson, C. Timothy Wehr, and Lori Tallman

Varian Instrument Group
2700 Mitchell Drive
Walnut Creek, CA 94598

INTRODUCTION

Steric exclusion chromatography (SEC) using aqueous mobile phases (traditionally referred to as gel filtration chromatography - GFC) is widely used for the separation and characterization of water-soluble organic molecules, especially biopolymers. GFC has been widely employed in biochemistry since crosslinked dextran gels were first introduced by Porath and Flodin in 1959.[1] Classical aqueous exclusion separations have utilized relatively soft, hydrophilic gels such as synthetic polyacrylamides, crosslinked polysaccharide dextrans (e.g., Sephadex) and agaroses (e.g., Sepharose, Bio-Gel A). These gels are characterized by relatively low compressive strength and must be operated at low pressures and flow velocities, which greatly limits their utility in HPLC.

Controlled-pore glasses and silica gel packings allow analysis at higher pressures and greatly reduced separation times compared to the soft gels traditionally used in aqueous exclusion chromatography; however, their active surface sites often result in adsorption effects. The slightly hydrophilic silica-based packings which contain bonded phases such as alkyl silanes with diol functionality reduce adsorption (i.e., chemically deactivate the silica gel) but often exhibit low pore volumes, thus reducing resolution.

Several new column packings for use in aqueous steric exclusion chromatography (SEC) have been developed in the past several years, utilizing both hydrophilic organic polymer gels and silica-based supports containing bonded phases.[2,3,4,5] Two types of high performance aqueous steric exclusion packings have recently been developed by Toyo Soda (Tokyo, Japan), one being a rigid, silica-based

packing containing a hydrophilic bonded phase (TSK Gel Type SW) and
the other a crosslinked hydrophilic, polymer-based, semi-rigid gel
(TSK Gel Type PW).[6,7,8]

In this work, prepacked columns containing these two support
materials are characterized from the standpoint of comparative chromatographic performance. Practical utility of these columns for
aqueous SEC separations of synthetic water-soluble polymers, biopolymers, and small water-soluble organic molecules are compared
and non-exclusion effects using amino acid probes are characterized
for Varian MicroPak TSK SW type and TSK Gel PW type columns.

EXPERIMENTAL

Chromatography was performed on Varian Model 5000 LC systems
equipped with a refractive index detector and a UV-50 variable
wavelength absorbance detector. Chromatographic separations were
carried out at 25°C using Varian MicroPak TSK SW and TSK Gel PW
type columns (7.5 mm x 30 cm) at a mobile phase flow rate of 1 ml/
min. Sample injection volumes were 100 μl using a Valco manual
loop injector. Solvents used were .01 $\underline{M}$ KH_2PO_4 (pH 6.8) for amino
acid probe samples, 0.1 $\underline{M}$ KH_2PO_4 + 0.1 $\underline{M}$ KCl (pH 6.8) for proteins,
and deionized water for polyethylene glycol standards.

Samples of commercial polyethylene glycol (PEG) standards were
obtained from Toyo Soda Mfg. Co. Ltd. (Tokyo, Japan) and Jefferson
Chemical Co. (Austin, Texas). Aqueous solutions 0.1% w/v were used
throughout this study. PEG standards above MW 10,000 were dissolved
in aqueous solutions containing 0.5% ethanol to aid dissolution and
retard chain scission of the standard.

Proteins and amino acid probe samples were obtained from Sigma
Chemical Co. (St. Louis, Missouri). Detection of proteins was at
280nm and detection of amino acids at 210nm in the UV.

RESULTS AND DISCUSSION

Characterization of MicroPak TSK SW and TSK Gel Type PW Columns

MicroPak TSK SW column packing is a rigid, hydrophilic, spher-
ical, and porous silica-based gel that contains a chemically bonded
phase. The surface of the packing is covered with hydroxyl groups.[9]
TSK Gel Type PW packing is a semi-rigid, hydrophilic, crosslinked
polymer-based gel containing the group $-CH_2CHOHCH_2O-$ as the main
constituent component.[10] The exact structure of these supports has
not been published.

A comparison of the characteristics of MicroPak TSK SW and TSK
Gel PW columns is shown in Table 1. Values of exclusion limits,
efficiencies, pore volumes, particle sizes, and pore sizes are

TABLE 1

CHARACTERISTICS OF MICROPAK TSK GEL

TYPE SW AND TSK GEL TYPE PW COLUMNS

Column Type	Particle Size *	Pore Size * (Å)	M.W. Exclusion Limit PEG	Protein	Theoretical Plates/Meter (N/M)	Pore Volume P
2000SW	10 ±2μ	130	20,000	100,000	21,000	0.92
3000SW	10 ±2μ	240	40,000	400,000	19,000	1.33
4000SW	13 ±3μ	450	200,000	1 million (est)	17,000	1.52
1000PW	10 ±2μ	-	1,000	-	16,000	0.89
2000PW	10 ±2μ	50	4,000	15,000	17,000	0.87
3000PW	13 ±2μ	200	50,000	450,000	15,000	0.83
4000PW	13 ±2μ	-	200,000	-	14,000	0.78
5000PW	17 ±2μ	1,000	1 mill.(est)	>600,000	13,000	0.98

Note: 1) Method for calculation of theoretical plates: Sample: 1% w/v Soln. ETHYLENE GLYCOL
                                                      Mobile Phase: 1 ml/min H_2O
                                                      Detector: R.I.
                                                      Loading: 100μl

      2) Pore Volume = $\frac{Vp-Ve}{Ve}$ where Vp = permeation volume
                                          Ve = exclusion volume

* Data supplied by Toyo Soda.

listed for each column type. MicroPak TSK SW type columns have
higher efficiencies and pore volumes (permeabilities) than the PW
type columns, and therefore should offer higher resolution. The
PW type columns cover a much larger molecular size range than the
SW columns due to the availability of a wider range of pore sizes.
The PW columns also allow operation over a wider pH range (1 to 13)
than the silica-based SW columns (2.5 to 7.5) (see reference 6).

Figure 1 displays a vanDeemter-type plot of plate height (HETP)
versus linear velocity for some SW and PW type columns. Plate
height increased with flow velocity and was found to taper off at
high flow velocities. In comparing the 3000PW to the 4000SW column,
it can be seen that the SW column has a higher efficiency at all
flow rates examined. The 3000SW column has a smaller particle size
packing (10 μ) than either the 3000PW or 4000SW column packing (13
μ) and the lower plate height at all flow velocities in part reflects
this fact. In practice, flow velocities of 0.04 to 0.06 cm/sec,
roughly corresponding to 0.8 to 1.3 ml/min mobile phase flow rate,
have been found to offer the best compromise between speed and
efficiency for both column types.

Polyethylene glycol (PEG) calibration curves for SW and PW
columns are shown in Figures 2 and 3. The slopes exhibited in the
linear region of the calibration curve for the SW and PW columns
are indicative of the resolution attainable with these columns.[11]

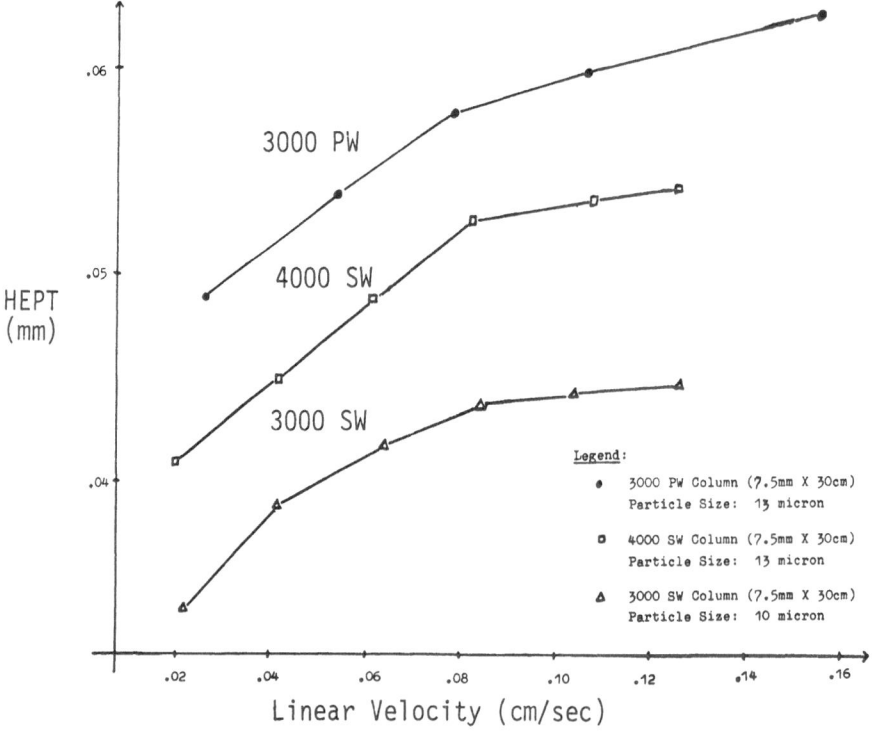

FIGURE 1.   Plot of Linear Velocity vs HETP

Comparative Performance of SW and PW Type Columns

Chromatographic column performance has been traditionally expressed in terms of the number of theoretical plates or efficiency:

$$N = 16 \left(\frac{V_R}{W} \right)^2 \tag{1}$$

where V_R is peak retention volume and W the peak width at baseline as measured by a peak triangulation technique.

The resolution (or separation efficiency) of a two-component mixture has also been used as a column performance parameter as described by the following equation:

$$R_S = \frac{2(V_{R2} - V_{R1})}{W_1 + W_2} \tag{2}$$

where V_{R1} and V_{R2} are the elution volumes of two solutes and W_1 and W_2 their respective peak widths.

The values of R_S and N calculated by these equations, however, are highly dependent on column dimensions and on the solutes chosen to characterize performance. Additionally, in steric exclusion chromatography, it would be very desirable to relate chromatographic

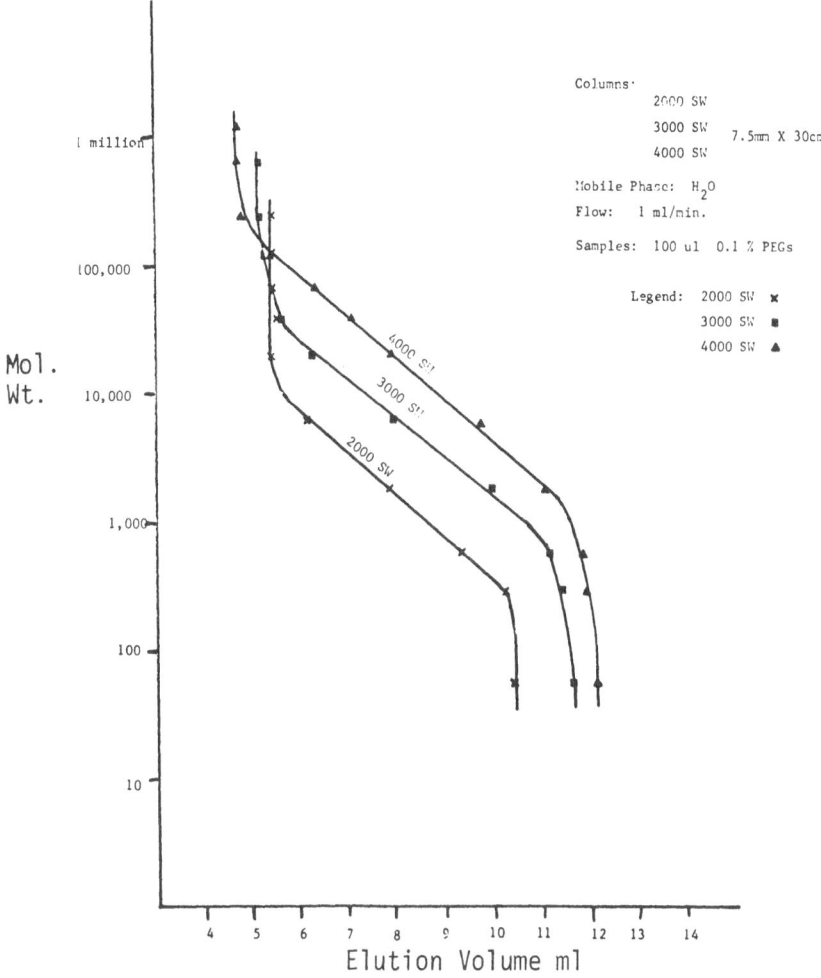

Columns: 2000 SW
3000 SW   7.5mm X 30cm
4000 SW

Mobile Phase: H_2O
Flow: 1 ml/min.

Samples: 100 ul 0.1 % PEGs

Legend: 2000 SW ✕
3000 SW ■
4000 SW ▲

FIGURE 2.   PEG Calibration Curve for MicroPak TSK SW

resolution to molecular weight since separation is based upon molecular size discrimination.

The concept of specific resolution, R_{SP}, was introduced by Bly who has shown that in the linear region of the calibration curve of log MW versus V_R for an exclusion column, resolution can be normal-ized and expressed as a function of molecular weights of a solute pair:[12]

$$R_{SP} = \frac{2(V_{R2} - V_{R1})}{W_1 + W_2} \cdot \frac{1}{\log (MW_1/MW_2)} \qquad (3)$$

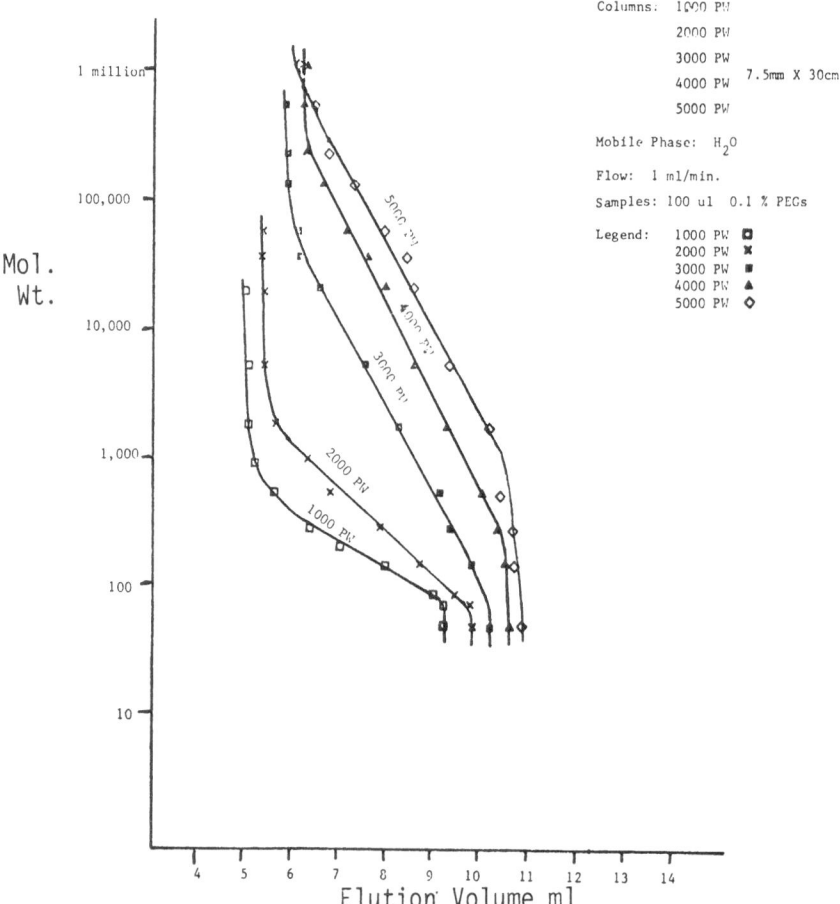

FIGURE 3.   PEG Calibration Curve for TSK PW

where MW_1 and MW_2 are the molecular weights of two solutes. For a pair of solutes with a decade difference in MW, equation 3 reduces to the expression for resolution. Additionally, outside the linear calibration range for a steric exclusion column, R_{SP} approaches zero.

Specific resolution, R_{SP}, is independent of the solute probes if the samples have very narrow molecular weight distribution (MWD).[13] Thus, R_{SP} is a more descriptive parameter for accurate performance comparison of steric exclusion column types. This parameter has been applied by Kirkland and Antle to the performance characterization of high performance steric exclusion packings in organic solvent systems.[14]

Specific resolution, R_{Sp}, values were calculated for a series of narrow MWD polyethylene glycol standards and protein standards with both SW and PW columns. The average molecular weight of a pair of PEG standards used to calculate R_{Sp} values can be defined as follows:

$$\text{Average MW} = \frac{MW_1 + MW_2}{2} \tag{4}$$

A plot of R_{Sp} versus Average MW defines the molecular weight range of optimum resolution for a steric exclusion column and provides a practical performance criteria for column selection and comparison in steric exclusion analysis. Such plots have been used by Kato et al. to characterize protein separations on TSK Type SW columns.[15]

Comparison of Small Pore Size SW and PW Columns for Small Molecule Analysis

Figure 4 displays a plot of specific resolution, R_{Sp}, versus Average MW for small pore size SW columns (2000SW, 3000SW) and PW columns (1000PW, 2000PW, 3000PW) using PEG standards. Several points are of interest:
1.   In comparing the 2000SW and 3000SW columns, little difference is seen in the specific resolution below MW 1000, while the 3000SW column displays higher resolution for larger solutes. Thus an ad-

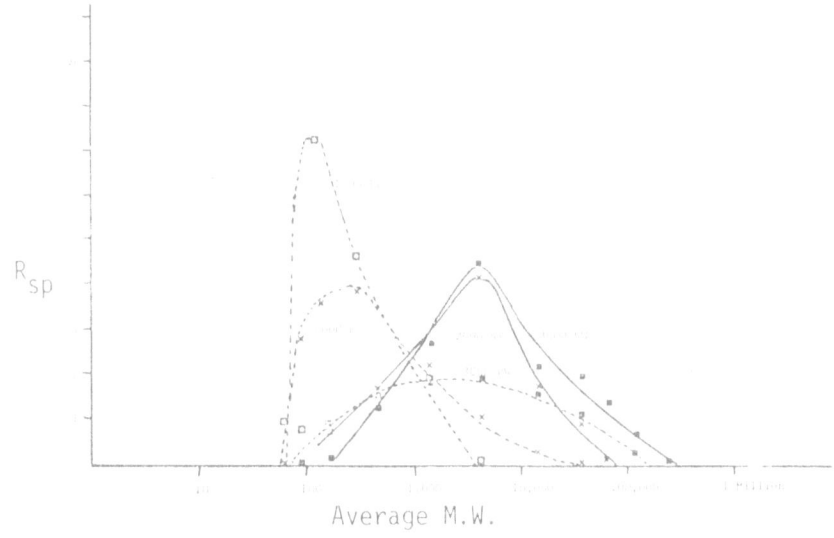

FIGURE 4.   Comparison of Specific Resolution for Smaller Pore Size
MicroPak TSK SW & PW Using PEG Stds.

vantage is gained by use of a 3000SW column over a 2000SW column. This fact has also been observed for protein separations using 2000SW and 3000SW columns (see reference 15);

2.   For low molecular weight samples (MW < 1000), the 1000PW and 2000PW columns exhibit very large specific resolution.   These columns are optimum for small molecule analysis;

3.   In comparing the 3000SW column to the 3000PW column, the 3000SW column exhibits higher specific resolution over most of the MW region (MW > 1000 and above) but the 3000PW exhibits higher specific resolution in the low MW region (MW < 1000).

Figures 5 and 6 show the analysis of PEG mixtures on a 3000SW and 3000PW column respectively.   These chromatograms visually illustrate the difference in specific resolution between the 3000SW and 3000PW columns as mentioned above.   The 3000SW column (Figure 5)

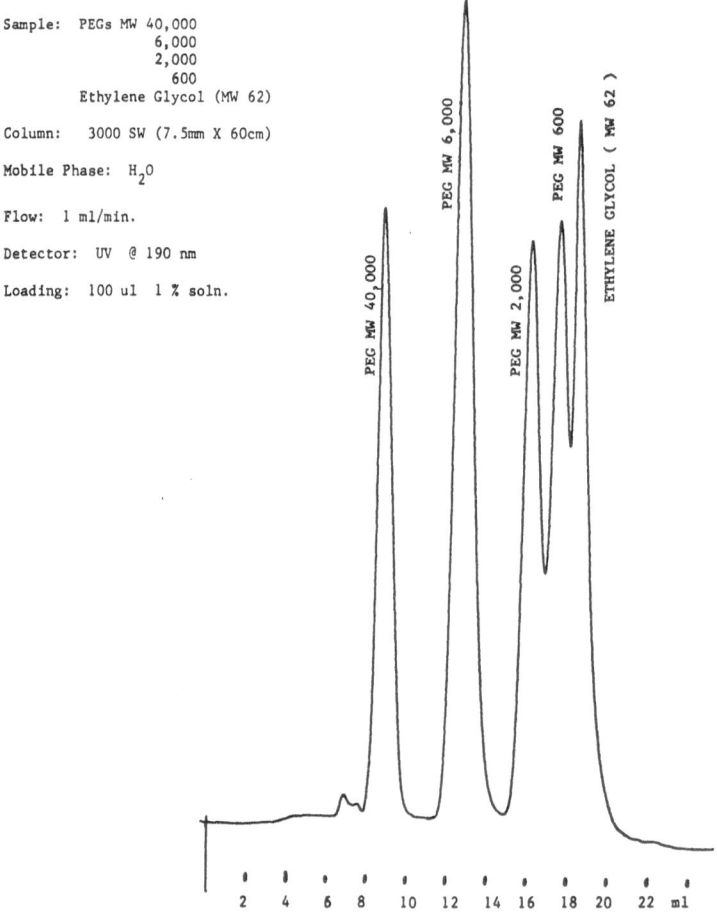

FIGURE 5.   PEG Stds. on MicroPak TSK 3000 SW

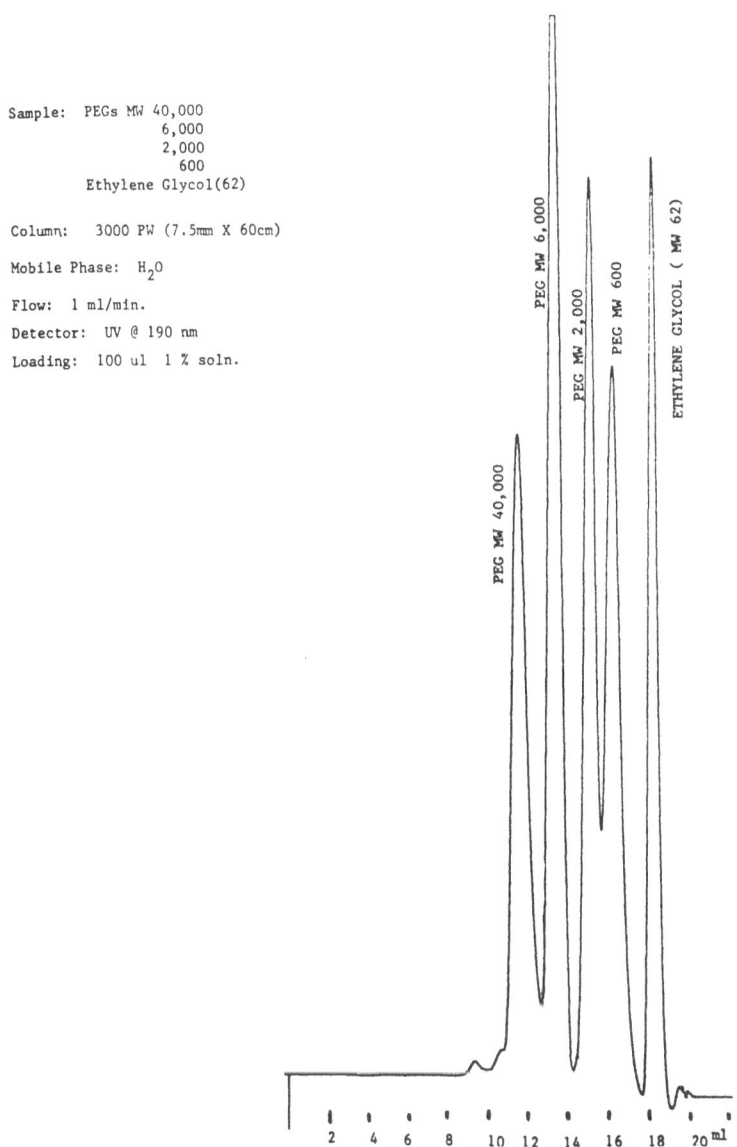

Sample: PEGs MW 40,000
                   6,000
                   2,000
                     600
         Ethylene Glycol(62)

Column: 3000 PW (7.5mm X 60cm)

Mobile Phase: H₂0

Flow: 1 ml/min.

Detector: UV @ 190 nm

Loading: 100 ul 1 % soln.

FIGURE 6.   PEG Stds. on TSK 3000 PW

exhibits higher resolution of the PEG samples MW 40,000 to 2,000,
while the 3000PW column (Figure 6) exhibits higher resolution of
PEG solutes of MW < 1000.   The very high resolution for small
molecules shown by the 1000PW column is depicted in the analysis of
several oligosaccharides shown in Figure 7.   Note that glucose and
xylose differ in MW by only 30 units.

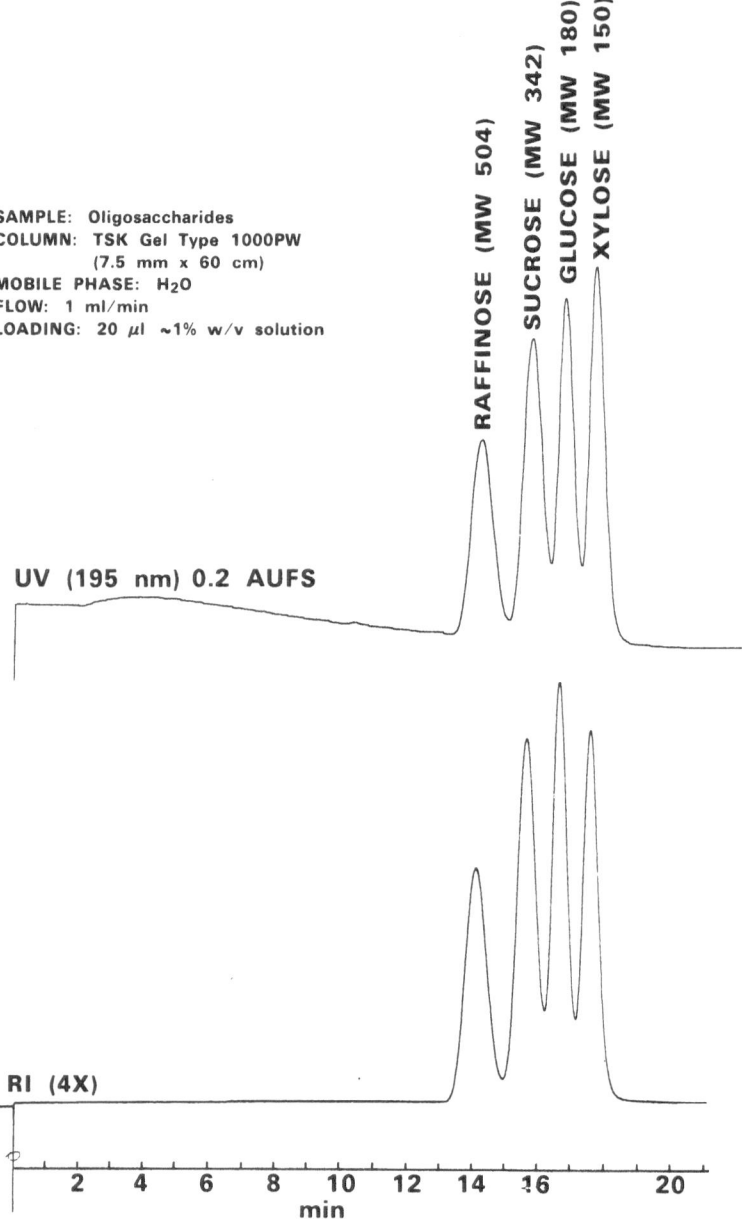

FIGURE 7.   Small molecule analysis on TSK type 1000 PW

Comparison of Large Pore Size SW and PW Columns for Water Soluble Polymer Analysis

A comparison of specific resolution curves for large pore size SW columns (4000SW) and PW columns (4000PW, 5000PW) using PEG standards is shown in Figure 8. Although the 4000SW column exhibits a higher specific resolution, its molecular weight range is more res-

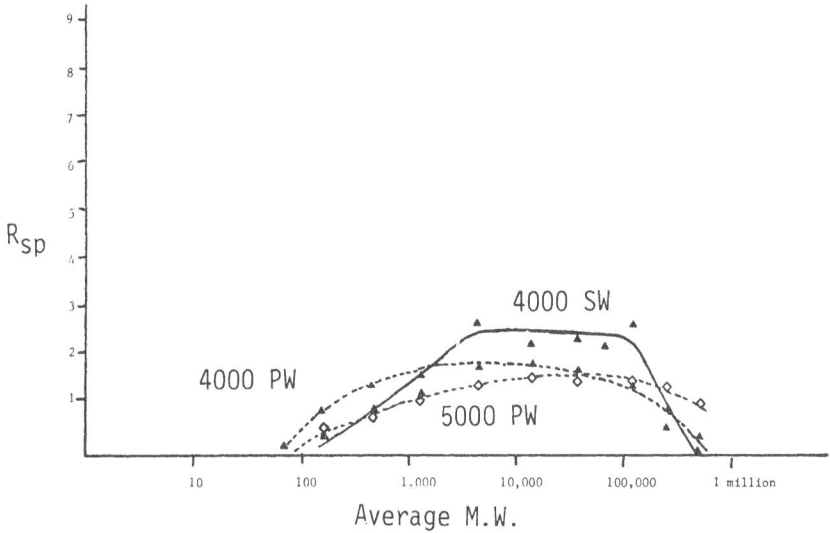

FIGURE 8.   Comparison of Specific Resolution Curves for Larger Pore
            Size MicroPak TSK SW & PW Using PEG Stds.

tricted than that of the 4000PW column in the low MW region and
5000PW column in the high MW region. For water soluble polymers of
MW > 200,000, the 5000PW column offers higher resolution.

    The influence of the mobile phase in aqueous steric exclusion
chromatography is particularly important because of its effect on
the conformation (and size) of the solute. The ionic strength of
the mobile phase is critical in eliminating electrostatic interactions between ionic groups on the column packing and ionic groups
on polymer samples (polyelectrolytes) that lead to attendant problems of ion exclusion, ion inclusion, ion exchange, and absorptive
effects.[16] For SW and PW type columns, ionic strengths of > 0.1
are preferred, and polyvalent anions seem to be more effective in
eliminating non-steric exclusion effects for most polymers. Polysaccharides, poly(vinyl alcohol), and poly(vinyl pyrollidone) polymers can be chromatographed on both SW and PW type columns using
low ionic strength mobile phases such as 0.02 $\underline{M}$ KH$_2$PO$_4$. Figures 9
and 10 show the analysis of several poly(vinyl alcohol) polymers
chromatographed respectively on a 4000SW and 5000PW column. Note
that the MW 96,000 PVOH polymer was partially excluded on the 4000SW
column.

    For very polar synthetic polymers such as polyacrylamide,
poly(acrylic acid), and polyethyleneimine, mobile phase ionic

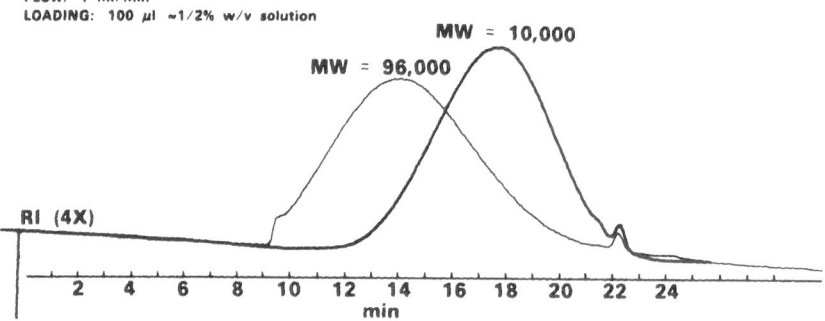

SAMPLE: Poly (vinyl alcohol) polymers
COLUMN: MicroPak TSK 4000SW
(7.5 mm x 60 cm)
MOBILE PHASE: 0.02M KH$_2$PO$_4$
FLOW: 1 ml/min
LOADING: 100 μl ~1/2% w/v solution

MW = 10,000

MW = 96,000

RI (4X)

FIGURE 9. Poly(vinyl alcohol) polymers on 4000 SW

strengths of > 0.4 are found to give satisfactory results for PW
type columns. The SW type columns exhibit adsorption effects, even
at these mobile phase ionic strengths; thus the PW type columns
give better performance for polar water-soluble polymer analysis.
Additionally, due to the limited pore sizes available with the SW
type columns, the PW type columns are better suited for analysis
of polymers with MW $\geq$ 100,000. Figure 11 displays the chromatograms
of poly(acrylic acid) polymers analyzed on a 5000PW column.

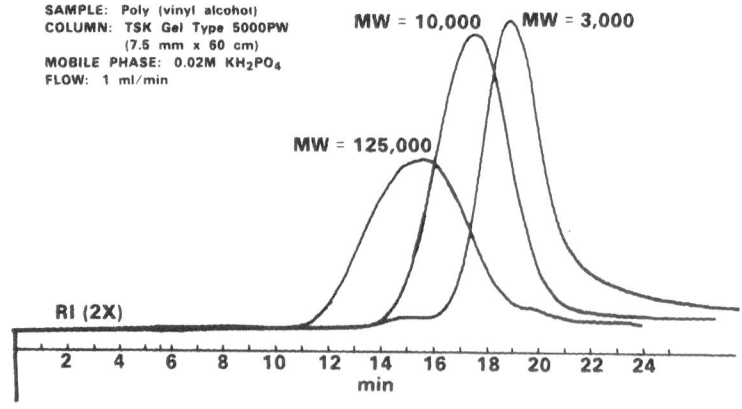

SAMPLE: Poly (vinyl alcohol)
COLUMN: TSK Gel Type 5000PW
(7.5 mm x 60 cm)
MOBILE PHASE: 0.02M KH$_2$PO$_4$
FLOW: 1 ml/min

MW = 10,000   MW = 3,000

MW = 125,000

RI (2X)

FIGURE 10. Poly(vinyl alcohol) polymers on 5000 PW

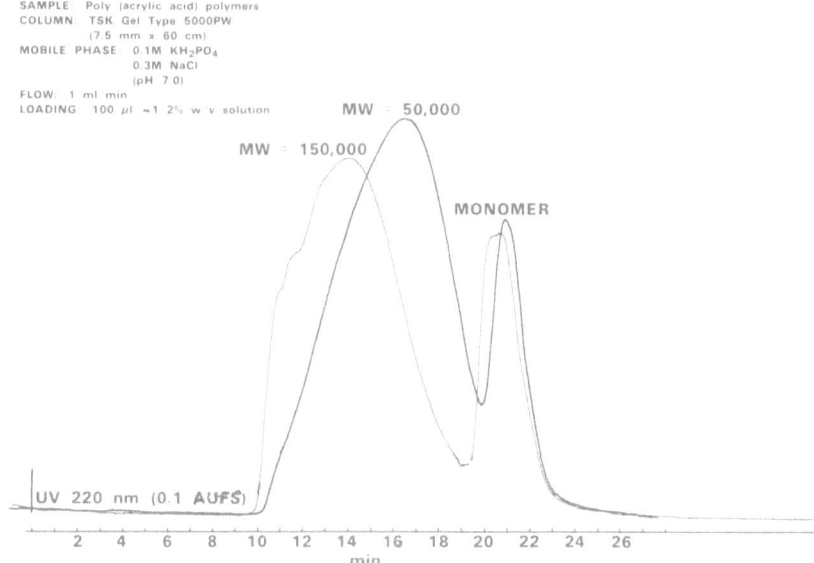

SAMPLE: Poly (acrylic acid) polymers
COLUMN: TSK Gel Type 5000PW
(7.5 mm x 60 cm)
MOBILE PHASE: 0.1M KH₂PO₄
0.3M NaCl
(pH 7.0)
FLOW: 1 ml/min
LOADING: 100 μl ~1.2% w/v solution

MW = 50,000

MW = 150,000

MONOMER

UV 220 nm (0.1 AUFS)

2   4   6   8   10   12   14   16   18   20   22   24   26
min

Figure 11.   Poly(acrylic acid) Polymers on 5000 PW Column

Comparative Performance of 3000SW and 3000PW Columns for Protein Analysis

High speed aqueous exclusion chromatography is a promising
tool in protein chemistry for semi-preparative isolation of proteins
as well as molecular weight characterization.[17]   Calibration curves
for 3000SW and 3000PW columns using protein standards are shown in
Figure 12.   These columns have been found to be the most useful of
the SW and PW type columns for protein separations.

The SW type columns exhibit little adsorptivity for proteins
as evidenced by nearly quantitative recoveries.[18]   The absorption
effects of proteins on 3000PW and 5000PW columns are slightly greater
than the SW type columns; however, recovery of proteins from these
PW type columns is > 80% in most cases.

The 3000SW column offers higher resolution for protein analysis
than the 3000PW column.   This is evident not only from the slope of
the calibration curves, but also from the comparison of specific
resolution curves shown in Figure 13.   As was the case for PEG standards, the 3000SW column provides higher resolution over most of the
MW region (MW > 1000) whereas the 3000PW column exhibits higher resolution of solutes of MW < 1000.   This fact is illustrated by the
chromatograms of thyroglobulin, albumin, ribonuclease A, and adeno-
sine shown in Figures 14 and 15.   Figure 14 displays the analysis

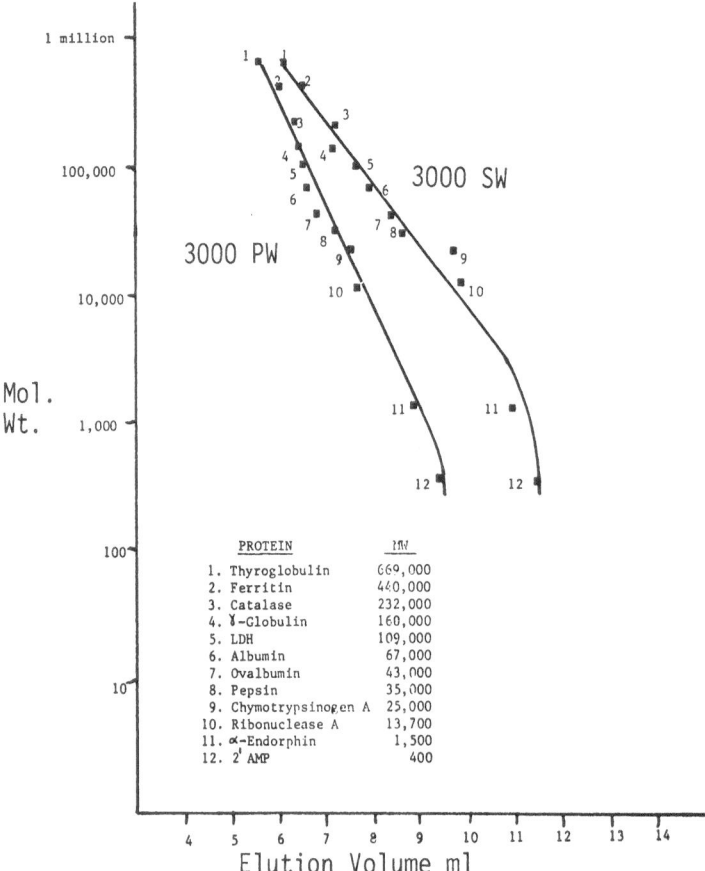

FIGURE 12.   Protein Calibration Curves for MicroPak TSK 3000 SW and
              3000 PW Columns

of these components on a 3000SW column and Figure 15 the analysis
on a 3000PW column. For thyroglobulin, albumin, and ribonuclease
A, higher resolution is achieved on the 3000SW column. However,
due to the wider pH range of operation possible with PW columns,
distinct advantages are offered by the PW columns for some biopoly-
mer separations.[19]

Characterization of Non-Exclusion Effects on SW and PW Columns

Non-exclusion effects resulting from solute-support interactions
can be broadly classified as arising from hydrophilic (dipole-dipole)
interactions and hydrophobic (Van der Waal dispersion force) interactions or partitioning.

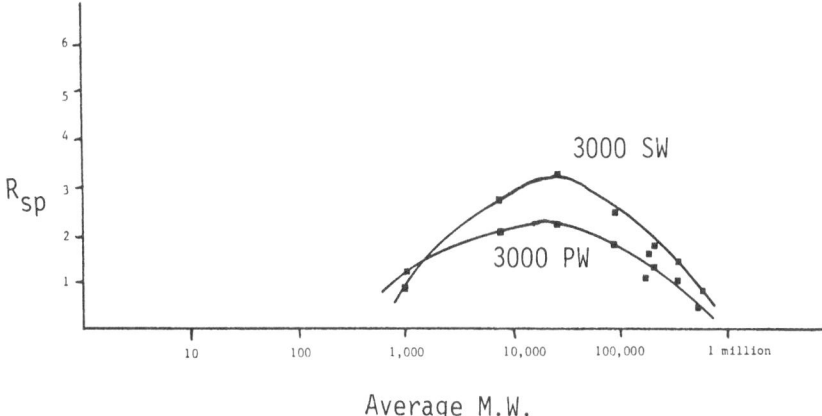

Average M.W.

FIGURE 13.   Comparison of Specific Resolution Curves for MicroPak
             TSK 3000 SW & 3000 PW Using Protein Stds.

An example of hydrophobic interactions on a 2000SW column can
be seen in Figure 16 which shows a chromatogram of xanthines analyzed
with a mobile phase of H_2O containing 5% MeOH (top) in which all
three components coeluted at the permeation volume as would be expected, and analyzed in a mobile phase of 0.2 $\underline{M}$ KH_2PO_4 in which
resolution of the three components occurred beyond the permeation
volume.

Figure 17 illustrates the effect of increasing mobile phase
ionic strength on the retention volume of xanthines. As ionic
strength increased, the K_D of all three xanthines increased (caffeine more so than the others). This is probably caused by a partitioning effect involving the organic bonded phase sometimes referred to as a "salting-in" effect. Caffeine, which contains an
extra methyl group, was slightly more affected by ionic strength
than the other two xanthines.

In order to more clearly define hydrophobic as well as other
non-exclusion interactions on SW and PW columns, a series of amino
acids were chosen as test probes. These compounds were chosen in
part due to a previously measured "hydrophobicity scale" defined
by Rekker[20] as shown in Figure 18. This figure shows the summation
of fragmental hydrophobicity constants for each amino acid. The
negative numbers represent hydrophilic amino acids. The more positive the summation of the fragmental constants, the more "hydro-

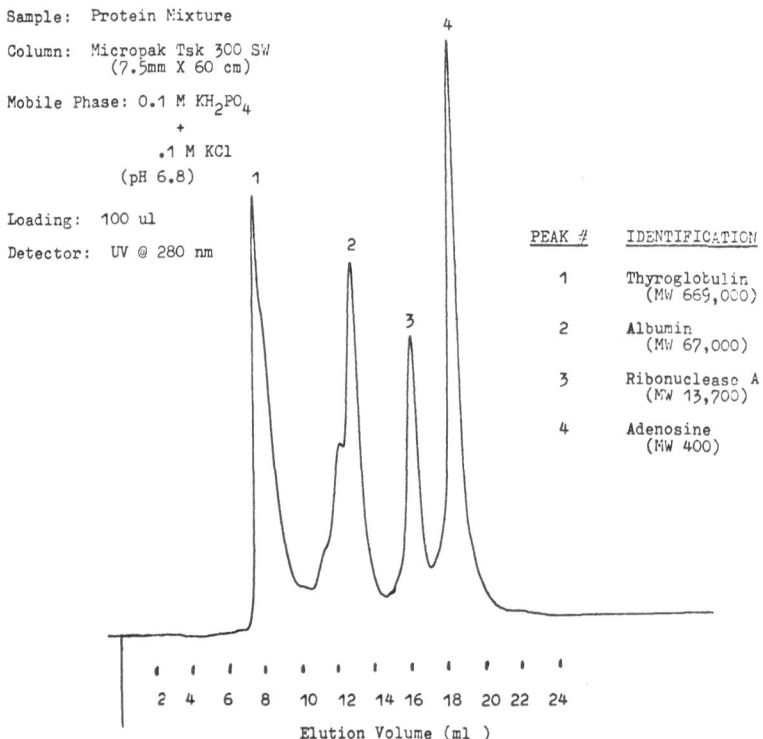

FIGURE 14.   Protein Stds. on MicroPak TSK 3000 SW

phobic" the amino acid. Tryptophan, phenylalanine, leucine, tyrosine, valine, and cysteine were chosen as test probes. To serve as controls, mono-, di-, tri-, and tetra-glycine (MW ∿ 100 to 500) were analyzed on each column to ensure differences in amino acid retention were not due to molecular size differences. Glycine oligomers coeluted on all columns tested except the 2000PW column which exhibited some separation. It should also be noted that of the amino acid probes used, tryptophan, phenylalanine, and tyrosine were aromatics, whereas leucine, valine, and cysteine were non-aromatic compounds.

The amino acid probes were analyzed on 2000PW, 3000PW, and 5000PW columns as well as 2000SW, 3000SW, and 4000SW columns with a mobile phase of 0.01 $\underline{M}$ KH$_2$PO$_4$ (pH 6.8). Results are displayed in graph form in Figure 19. Summation of the hydrophobic fragmental constant is plotted versus k' for the amino acid probes. Several points are noteworthy:
1.   A non-linear relationship exists between k' and hydrophobicity

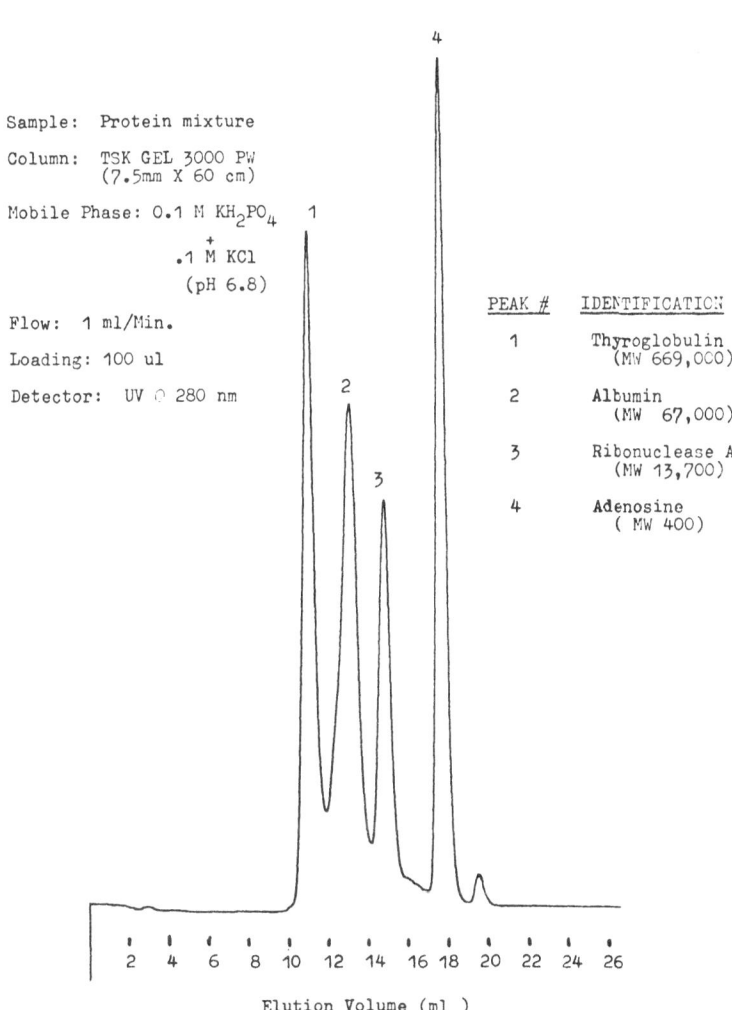

Sample: Protein mixture

Column: TSK GEL 3000 PW
         (7.5mm X 60 cm)

Mobile Phase: 0.1 M KH$_2$PO$_4$
              +
              .1 M KCl
              (pH 6.8)

Flow: 1 ml/Min.

Loading: 100 ul

Detector: UV @ 280 nm

PEAK #	IDENTIFICATION
1	Thyroglobulin (MW 669,000)
2	Albumin (MW 67,000)
3	Ribonuclease A (MW 13,700)
4	Adenosine (MW 400)

Elution Volume (ml)

FIGURE 15. Protein Stds. on TSK 3000 PW

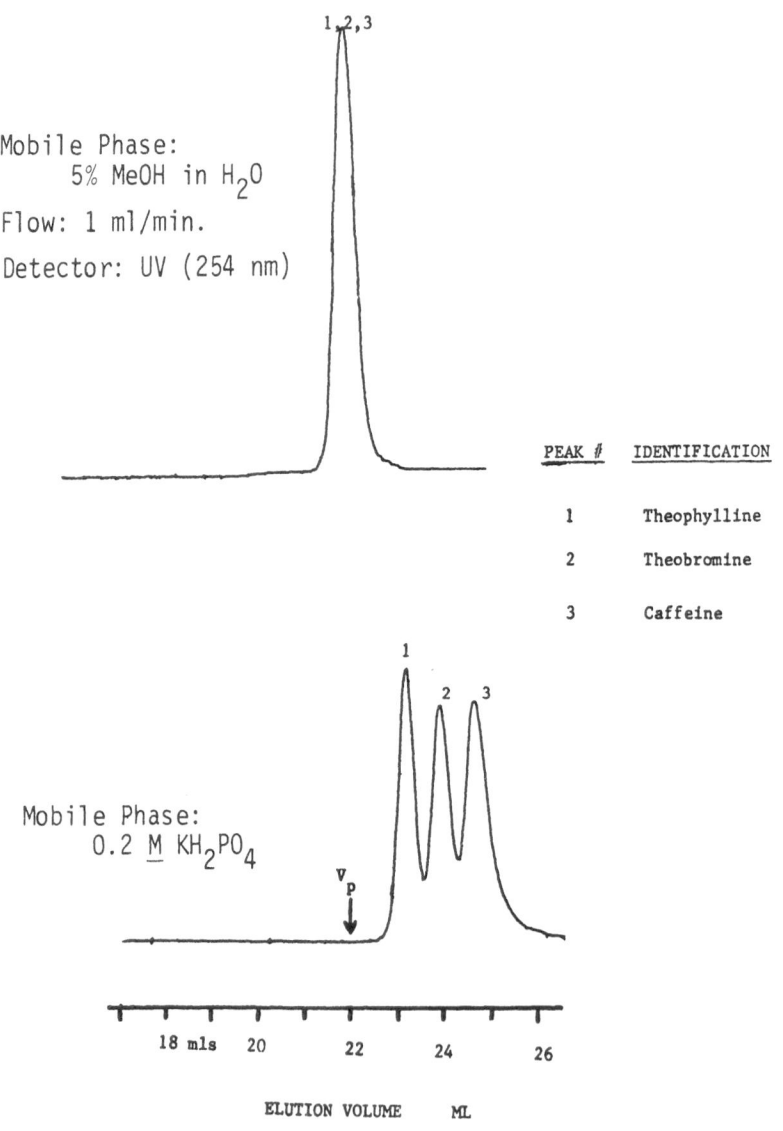

FIGURE 16.   Hydrophobic Interaction of Xanthines on MicroPak TSK
             2000 SW

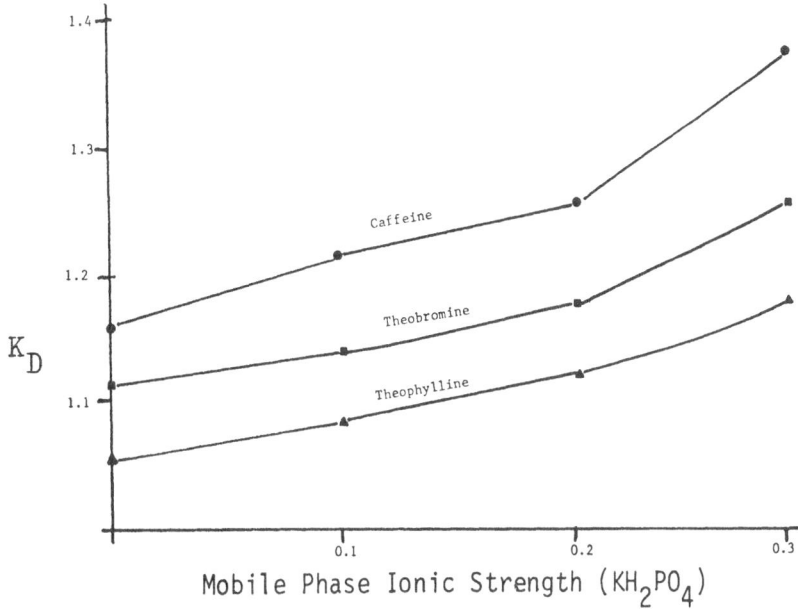

FIGURE 17.   Plot of K_D of xanthines versus ionic strength on MicroPak
TSK 2000 SW

Amino Acid	Σf	pK$_a$ of Basic Side Chains
Tryptophan	2.31	(Aromatic)
Phenylalanine	2.24	"
Leucine	1.99	(Non-Aromatic)
Isoleucine	1.99	"
Tryosine	1.70	(Aromatic)
Valine	1.46	(Non-Aromatic)
Cystine	1.11	
Methionine	1.08	
Proline	1.01	
Cysteine	0.93	(Non-Aromatic)
Alanine	0.53	
Lysine	0.52	10.53
Glycine	0.00	
Aspartic Acid	-0.02	
Histidine	-0.23	6.00
Threonine	-0.26	
Serine	-0.56	
Asparagine	-1.05	
Glutamic Acid	-1.09	

FIGURE 18.   Hydrophobic fragmental constants of the common amino acids.

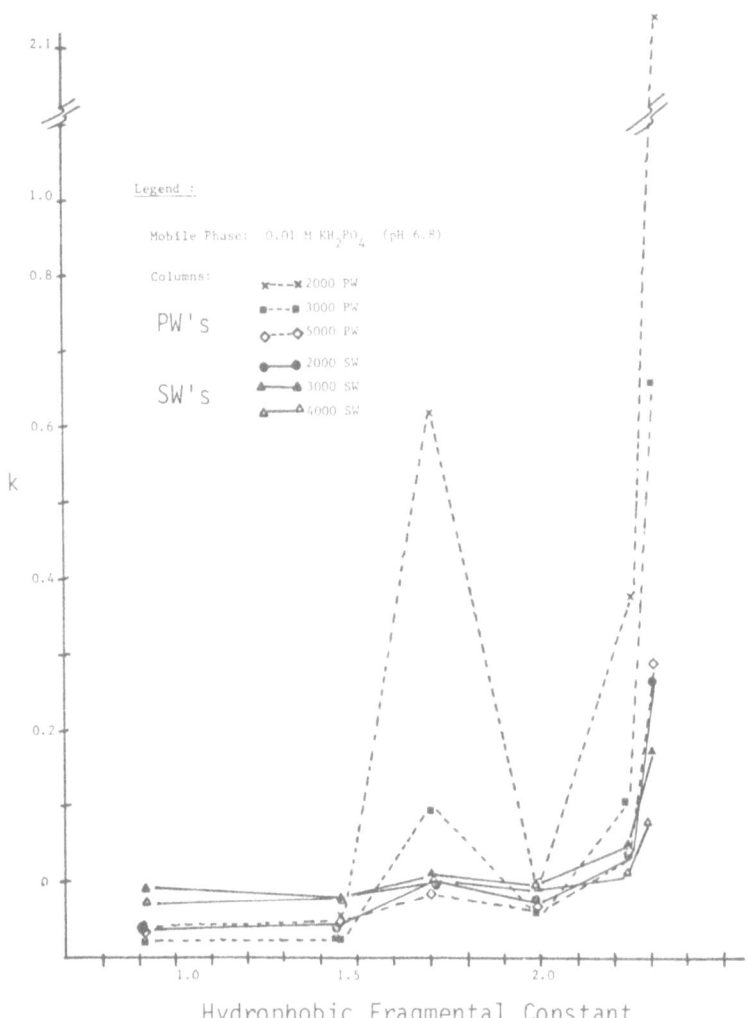

FIGURE 19.   Comparison of hydrophobic interactions of Micropak TSK
             SW & PW using amino acid probes

as measured by Σf;
2.   The PW columns seem very sensitive to aromatic compounds which
exhibit greater retention (higher k') than the non-aromatic compounds. This is clearly demonstrated for tyrosine (Σf = 1.7);
3.   The sensitivity to aromatic compounds of the PW columns greatly
increases with decreasing pore size.

     Ion exchange and ion exclusion interactions were evaluated
using charged or zwitterionic amino acid test probes of similar
molecular weights:

    Pair 1. Methionine   - zwitterion, MW 149
             Lysine   - net charge (+1), MW 146

    Pair 2. Threonine   - zwitterion, MW 119
             Aspartic Acid- net charge (-1), MW 133

These pairs of hydrophilic amino acid test probes were analyzed on
PW and SW columns in mobile phases of increasing ionic strength.
Non-steric exclusion effects as measured by retention volume (k')
as a function of ionic strength were negligible on both SW and PW
columns using these amino acid solute probes.

CONCLUSION

    MicroPak TSK SW type columns have a more limited MW separation
range than the TSK Gel PW type columns, although SW type columns
exhibit higher resolution. Due to both limited pore size range and
adsorption effects of polar, synthetic water-soluble polymers on SW
columns, PW columns are recommended for the analysis of synthetic
water-soluble polymers.

    The SW type columns offer higher resolution than PW type columns
for protein analysis (3000SW most useful). However, small pore size
PW columns (1000PW and 2000PW) are best suited for small molecule
analysis (MW 100 to 1,000).

    Some hydrophobic interaction occurs with both SW and PW columns.
The PW columns seem much more sensitive to aromatic compounds than
SW columns, and the effect seems to increase with decreasing pore
size.

REFERENCES

1.   J. Porath and P. Flodin, Nature 183 (1959), p. 1657.
2.   M. Vondruska, M. Sudrich, and M. Mladek, J. Chromatogr. 116
     (1976), p. 457.
3.   F. E. Regnier and R. Noel, J. Chromatog. Sci. 14 (1976), p.
     316.
4.   R. V. Vivilecchia, B. G. Lightbody, N. Z. Thimot, and H. M.
     Quinn, J. Chromatog. Sci. 15 (1977), p. 424.
5.   N. Becker and K. K. Unger, Chromatographia 12 (1979), p. 539.
6.   "SW Type Technical Data" and "PW Type Technical Data", Toyo
     Soda Mfg. Co. Ltd., Tokyo, Japan, 1978.
7.   K. Fukano, k. Komiya, H. Sasaki, and T. Hashimoto, J.
     Chromatogr. 166 (1978), p. 47.
8.   Y. Kato, K. Komiya, A. Sasaki, and T. Hashimoto, J. Chromatogr.
     193 (1980), p. 311.
9.   Toyo Soda Mfg. Co. Ltd., Tokyo Japan: "SW Type Technical Data"
     (1978).

10. T. Hashimoto, N. Sasaki, M. Aiura, and Y. Kato, J. Polymer
     Sci: Polymer Physics Edition 16 (1978), p. 1789.
11. A. J. deVries, M. LePage, R. Beau, and C. L. Guillemin, Anal.
     Chem. 39 (1967), p. 935.
12. D. D. Bly, J. Polymer Sci., Part 6, 21 (1968), p. 13.
13. W. W. Yau, J. J. Kirkland, D. D. Bly, and N. J. Stoklosa, J.
     Chromatogr. 125 (1976), p. 219.
14. J. J. Kirkland and P. E. Antle, J. Chromatog. Sci. 15 (1977),
     p. 137.
15. Y. Kato, M. Komiyo, H. Sasaki, and T. Hashimoto, J. Chromatogr.
     190 (1980), p. 297.
16. B. Stenlund, J. C. Giddings, E. Grushka, J. Cazes, and P.
     Brown, Advances in Chromatography, vol. 14, Marcel Dekker (1976),
     p. 37.
17. C. T. Wehr and S. R. Abbott, J. Chromatogr. 185 (1979), p. 453.
18. C. T. Wehr, F. E. Klink, and J. Robinson, "Recovery of Proteins
     from High Speed Steric Exclusion Columns", Applications Note
     #102, Varian Instrument Group, Walnut Creek, CA 94598. See
     also K. Fukano, K. Komiya, H. Sasaki, and T. Hashimoto, J.
     Chromatogr. 166 (1978), p. 47.
19. T. Hashimoto, H. Sasaki, M. Aiura, and Y. Kato, J. Chromatogr.
     160 (1978), p. 301.
20. R. F. Rekker, The Hydrophobic Fragmental Constant, Elsevier,
     Amsterdam, New York, 1977, p. 301.

EFFLUENT FREE ELECTROLYTIC REGENERATION

OF ION-EXCHANGE RESINS

H. Strathmann and K. Kock

Froschungsinstitut Berghof GmbH
D - 7400 Tübingen, Postfach 1523, Germany

INTRODUCTION

The efficient and economic removal of ionic species from industrial effluents is a problem which is
gaining increasing significance in today's waste water
treatment systems. Reverse osmosis and electrodialysis
have successfully been used in recent years for this
task. For solutions containing ionic components in
relatively low concentrations (less than 100 ppm), a
conventional ion-exchange technique is preferred today
for economic reasons. However, the regeneration of the
exhausted resin leads to an effluent which contains a
multitude of the stoichiometrically required quantities
of the ions used for the regeneration process, mixed
with the components removed from the original feed
solution. With valuable or highly toxic materials such
as certain heavy metal ions it is desirable to recover these ions in as concentrated a form as possible
and with a minimum of impurities.

A virtually effluent-free regeneration of a
charged ion-exchange resin with the extracted metal
in solid form can in some cases be achieved by a procedure referred to as electrolytical or electrodialytical regeneration[1]. Although the process has been
described in several patents[2] and in the literature
over 20 years ago, it has gained up until now only
little technical significance. With the increasing
costs of raw materials and pollution problems caused
by industrial waste waters, however, electrodialytic

regeneration might again become of interest. The
electrolytic regeneration of ion-exchange resins is of
special interest in the treatment of rinse cycle effluents from the electroplating industry which often
contain toxic or valuable heavy metal ions, such as
Cu^{++}, Cd^{++}, Pb^{++}. etc. in relatively low concentrations.
In this paper, the removal of Pb^{++}-ions obtained from
effluents of a lead battery production line and Cu^{++}-
ions obtained from the rinse cycle of an electroplating
process by ion-exchange and electrolytic regeneration
is discussed.

THE PRINCIPLES OF THE PROCESS

The principle of electrolytic regeneration is shown
schematically in Figure 1. A cation-exchange resin
which is charged with metal-ions is placed between
two cation-exchange membranes and two electrodes. The
electrode compartments contain electrolyte solutions
to provide the necessary conductivity. By applying a
direct current, H^{+}-ions generated at the anode migrate
through the cation-exchange membrane into the resin and
replace metal-ions which migrate through the opposite
membrane towards the cathode where they are electroche-
mically reduced and precipitated as solid metal. The
process can be continued until all metal-ions in the
resin are replaced by H^{+}-ions and precipitated at the
cathode. Ideally, the electric current to the cathode
is carried by the metal-ions yielding a current utilization for the regeneration process of 100 %. In practice,
however, this is never the case for reasons which will
be discussed later.

The main difference between electrodialytical and
conventional regeneration of an ion exchange resin is
that in conventional regeneration, H^{+}-ions are exchanged
from an aqueous solution into a swollen gel under the
driving force of a chemical potential gradient, while
in electrodialytical generation the H^{+}-ions are forced
into the resin under an electrical potential gradient,
leading to a virtually effluent-free regeneration where
the lead is obtained in a solid form precipitated at
the cathode.

EXPERIMENTAL PROCEDURES

The tests were carried out in a laboratory cell
which consisted of two compartments separated by an
ion-exchange column packed between two cation exchange

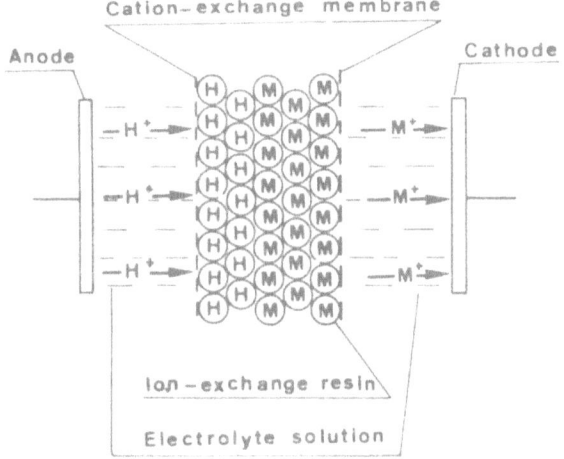

Fig. 1.   Schematic diagram of the electrolytic
           regeneration of a metal-ion charged
           cation-exchange resin

membranes as indicated in Figure 2. The membrane areas
facing the electrodes were 25 cm² each. The total volume
of the ion-exchange column was 20 cm³. The volumes of
the electrode chambers were 40 ml each. The anode consisted of a platinized titanium screen and the cathode
of a stainless steel or lead plate. The cation-exchange
resin was used either in the hydrogen form or converted
into the sodium form with a solution of $NaNO_3$. The
resin was charged by passing a solution containing
the metal ions through the column. The metal content
of the ion-exchange resin was determined from a mass-
balance by measurements of the volumes and concentra-

tions of the feed solution and the effluent. The feed
solutions were copper sulfate and lead acetate. The
concentrations were between 1000 and 5000 ppm. Two
different ion-exchange resins were used during the tests.
One resin, the Permutit RS-20^r, is a sulfonated polystyrene crosslinked with 2% DVB. This resin is weakly
cross-linked and the H$^+$-ion dissociation is high in a
pH range from 0 to 14.

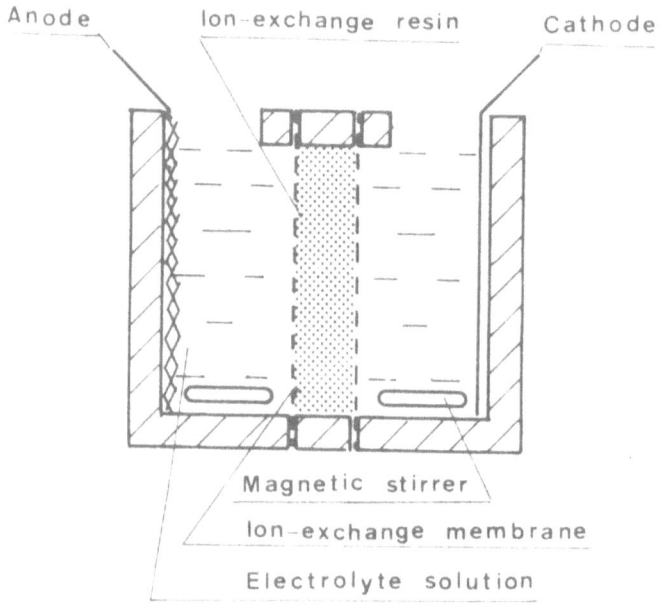

Fig. 2.   Cross-section of the test cell used for
           the electrolytical regeneration of a
           cation-exchange resin

The second resin, the Permutit C-67^r, has a microporous
structure and carries carboxylic groups; the H$^+$-ion
dissociation is therefore low, especially in the low
pH range. The electrode chambers were filled with an
aqueous solution of 2% formic acid (CH$_3$COOH) and 2%
sulfamic acid (NH$_2$SO$_3$H) when the lead charged resin
was regenerated and with 2% sulfuric acid when the
copper charged resin was regenerated. The total current

was adjusted to 0.1 A giving a current density through
the resin of 0.004 A/cm². During the tests, the resins
were rinsed with deionized water or, in some cases, with
a sodium nitrate solution. The voltage drop between the
electrodes was between 5 and 50 V depending on the
resin and the degree of regeneration. The current and
the voltage drop are continuously monitored while the
metal precipitation is determined by measuring the
weight increase of the cathode at certain time intervals.

TEST RESULTS

    The test results are summarized in the graphs in
Figures 3 to 12. In Figures 3, 5, and 7, the amount
of lead precipitated on the cathode is plotted as a
function of the total current which has been passed
through the apparatus, the solid lines indicating the
theoretical value that would be obtained when the
current utilization would be 100%. The different curves

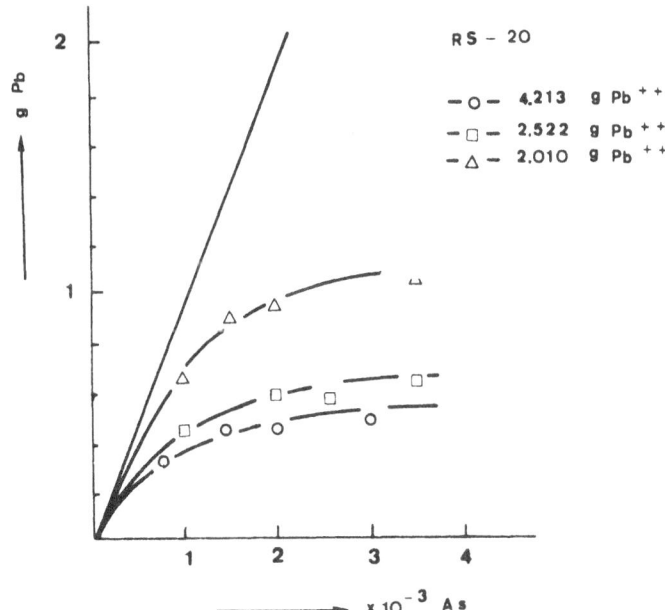

Fig. 8.   Lead precipitated on the cathode during
               the electrolytical regeneration of lead
               charged Permutit RS-20 cation-exchange
               resins as a function of the current
               passed through the test cell

show the results obtained with different degrees of
resin loading. The numbers RS-20 and C-67 indicate
the resins used during the tests. The Figures 4,6, and
8 show similar results, however, here the ratio of the
precipitated lead to the amount of lead originally in
the resin is plotted versus the total amount of current
passed through the test cell. The results of Figures
5 and 6, respectively, were obtained by adding about
2% NaNO$_3$ to the solution in the anode chamber. In
Figures 9 to 12, the results obtained by regenerating a
copper charged resin are correspondingly plotted.

DISCUSSION OF THE TEST RESULTS

        The test results indicate that the electrolytical
regeneration of a lead or copper charged cation-exchange
resin is, in principle, possible. The current utilization, however, is always significantly lower than

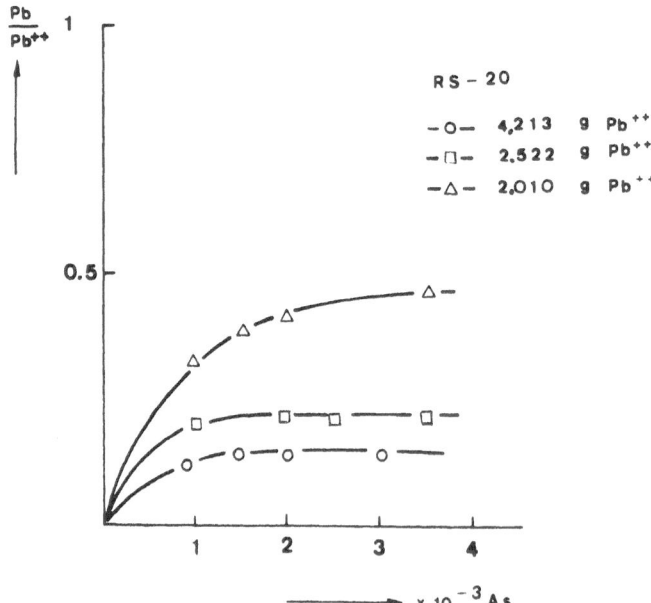

Fig. 4.   Degree of regeneration of lead charged Permutit
          RS-20 cation-exchange resins as a function of
          the current passed through the test cell.

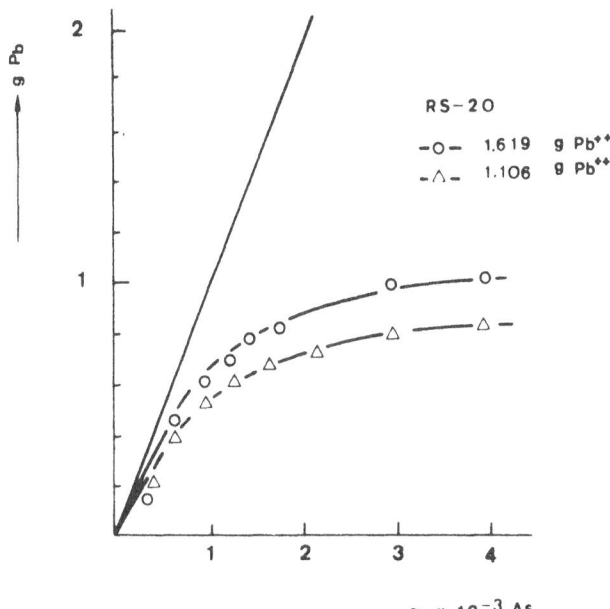

Fig. 5. Lead precipitated on the cathode during the elec-
         trolytical regeneration of lead charged Permutit
         RS-20 cation-exchange resins as a function of the
         electric current passed through the test cell.

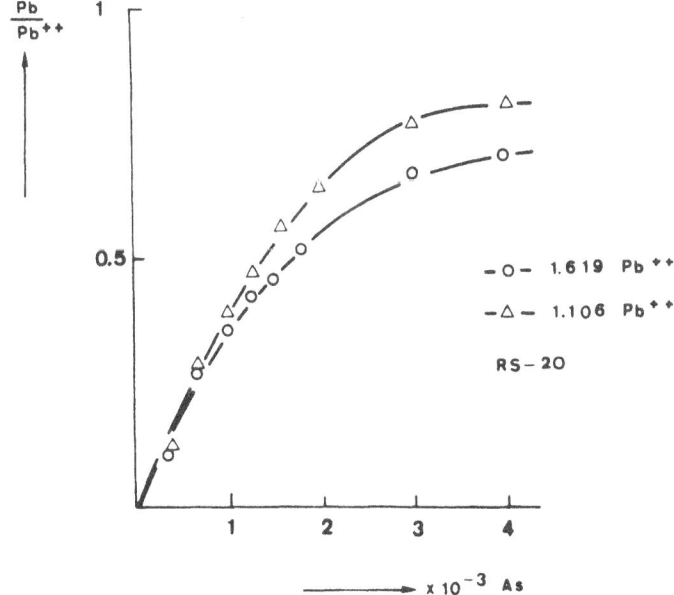

Fig. 6. Degree of regeneration of lead charged Permutit
         RS-20 cation-exchange resins as a function of the
         electric current passed through the test cell.

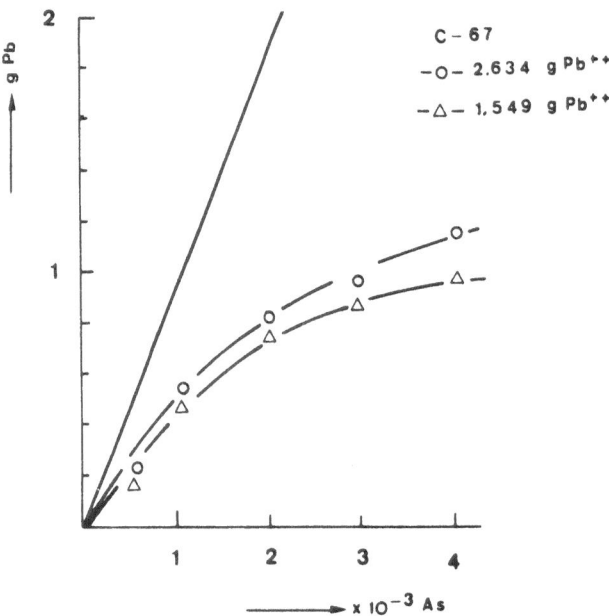

Fig. 7.   Lead precipitated on the cathode during the elec-
           trolytical regeneration of lead charged Permutit
           C-67 cation exchange resins as a function of the
           electric current passed through the test cell.

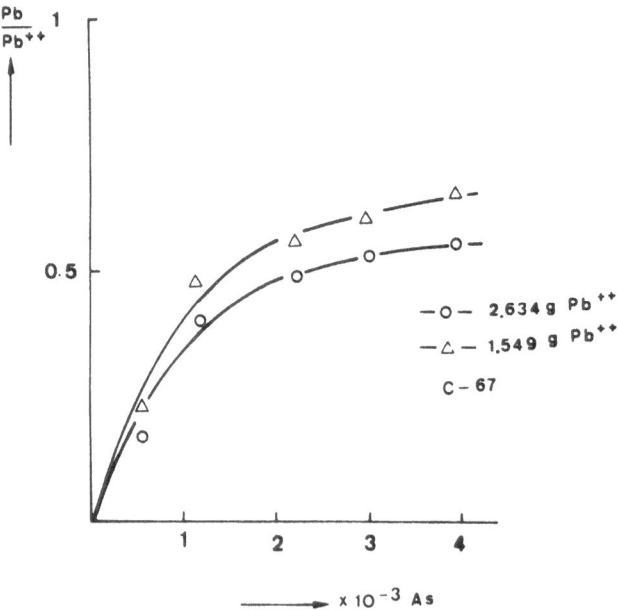

Fig. 8.   Degree of regeneration of lead charged Permutit
           C-67 cation-exchange resins as a function of the
           electric current passed through the test cell.

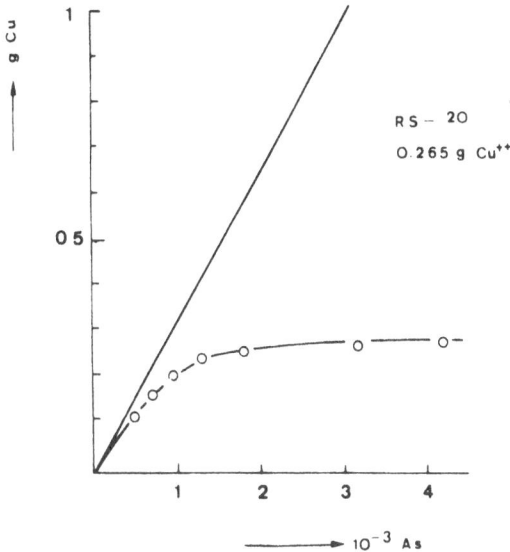

Fig. 9.   Copper precipitated on the cathode during
the electrolytical regeneration of a copper
charged Permutit RS-20 cation-exchange resin
as a function of the electric current passed
through the test cell

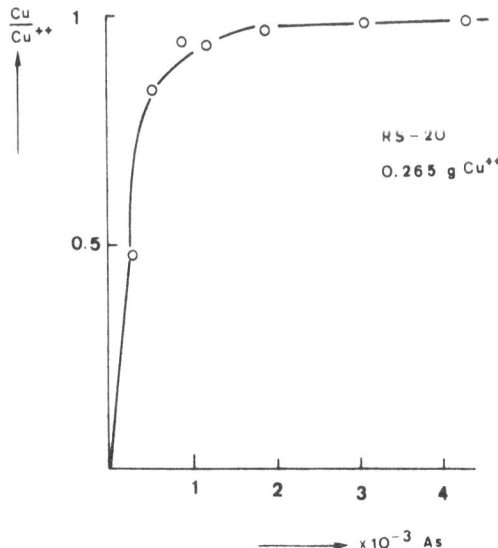

Fig. 10.   Degree of regeneration of a copper charged Permutit C-67 cation-exchange resin as a function
of the electric current passed through the cell

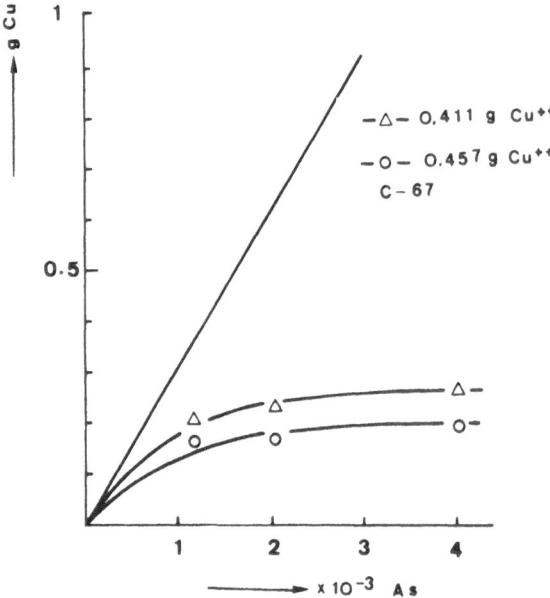

Fig. 11.   Copper precipitated on the cathode during the
            electrolytical regeneration of copper charged
            Permutit C-67 cation exchange resins as a
            function of the electric current passed
            through the test cell.

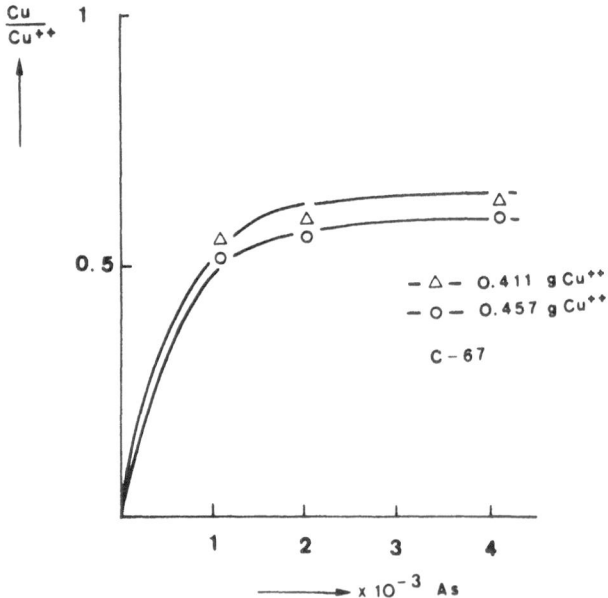

Fig. 12.   Degree of regeneration of copper charged Permutit
            C-67 cation-exchange resins as a function of the
            electric current passed through the test cell.

theoretically calculated. Furthermore, it decreases
with the degree of regeneration and reaches extremely
low values when a complete regeneration of the resin is
required. Since the current utilization determines to
a large extent the economics of the process, a high
current utilization is essential. The test results
shown in Figure 3 and 4 were obtained with resin RS-20
in the H^+ form. The current utilization was very poor.
The maximum regeneration that could be achieved with a
reasonable current utilization was 20 to 30%. Apparently,
the main problem associated with the electrolytic regeneration of cation-exchange resins is the different
mobilities of hydrogen- and metal-ions in the resin.
Since the mobility of the H^+-ions is much higher than
that of the Pb^{++} of Cu^{++}-ions, a "breakthrough" of H^+-ions
may occur as indicated in Figure 13. Here, the movement
of the regeneration front through the resin is shown
schematically. In a), the regeneration is just beginning
and the current from the anode to the resin is carried by
H^+-ions and from the resin to the cathode by the metal-
ions. The current density through the resin is identical
over the entire cross-section of the resin as indicated
by the size of the arrows. In b), a distortion in the
regeneration front is indicated and because of the higher
mobility of the H^+-ions, more current will pass through
the lower part of the column. The effect is autocatalytic
and will increase until a breakthrough of H^+-ions is
obtained as indicated in c). When this occurs, the
current will be nearly completely carried by H^+-ions
and the current utilization of the regeneration procedure
is drastically affected. To obtain a high degree of
regeneration at a reasonable current utilization, either
a distortion of the regeneration front has to be avoided
or the mobility of the H^+-ions in the resin has to be
decreased. Two attempts were made to affect the mobility
of the H^+-ions. In the first case, $NaNO_3$ was added to
the electrolyte solution in the anode compartment and
in the second case, the strong acid ion-exchange resin
was replaced by a resin with a low degree of H^+-ion
dissociation. In both cases, the current utilization
could be significantly improved, as the results in
Figures 5 and 8 indicate. The addition of $NaNO_3$ to the
anode solution apparently leads to a three step regeneration procedure. First, the lead is exchanged against
the Na^+-ions which then are replaced by H^+-ions. The
weakly dissociated ion-exchange resin leads to a significantly lower mobility of the H^+-ions in the resin.
Unfortunately, this is associated with a drastic increase
in the resistivity of the resin, leading to an increasing
energy consumption. Only when the resin was continuously

rinsed with a NaNO$_3$ solution could the resistivity be
lowered to acceptable values. The same test performed
with a CuSO$_4$ solution showed similar results, as can be
seen in Figures 9 to 12. The current utilization in
regeneration of a Cu^{++}-ion charged resin was, however,
slightly better than in the regeneration of the Pb^{++}-ion
charged resin. The precipitation of the copper was quite
clean, as a solid metal, while the precipitation of the
lead was spongy. In Figure 14, the energy required to
remove one equivalent of metal ion, i.e. in the case of
lead, 103.6 g and in the case of copper, 31 .77 g, from
the ion exchange resin and precipitate it on the cathode
is shown as a function of the voltage drop between the

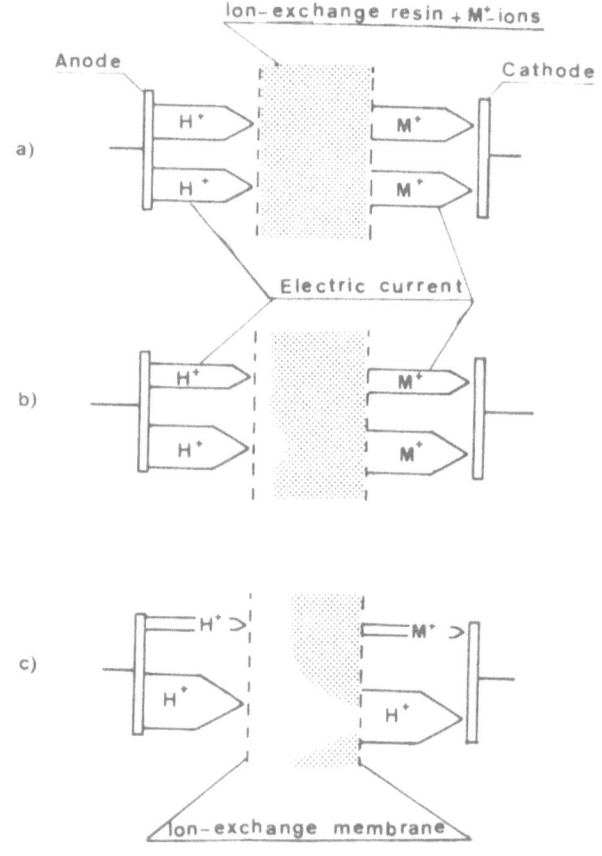

Fig. 13.   Schematic diagram indicating the movement
              of the regeneration front through a metal-
              ion charged cation-exchange resin and the
              breakthrough of the H$^+$-ions

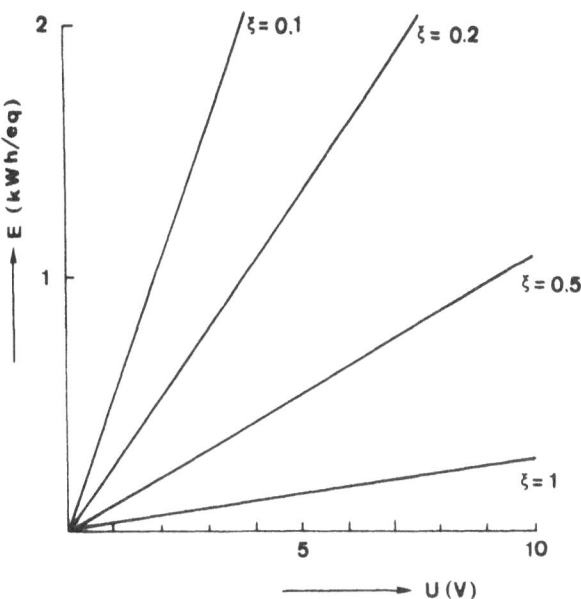

Fig. 14. Energy requirements for the electrolytic regen-
          eration of cation-exchange resins at various
          degrees of current utilization as a function
          of the voltage drop between the electrodes.

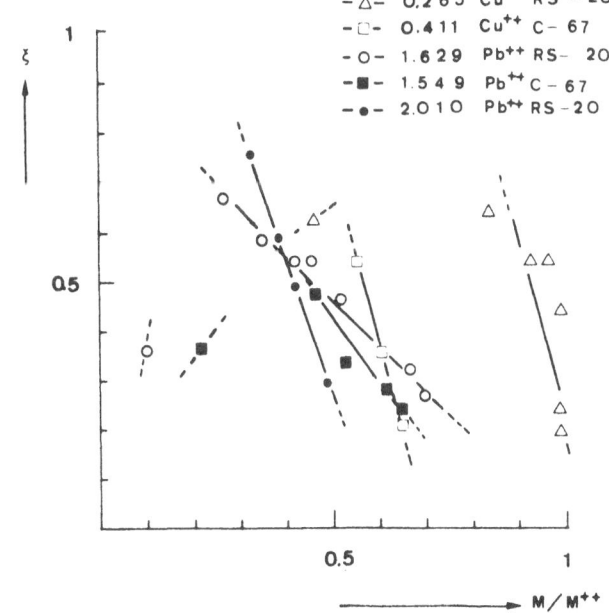

Fig. 15. Current utilization of various regeneration
          tests as a function of the degree of resin
          regeneration.

electrodes for various degrees of current utilization.
The graph indicates that, for the electrolytic regeneration process to be economical, the voltage drop should
be as low as possible and the current utilization as
high as possible. In the case of the recovery of lead
and copper, a current utilization of 50% and a voltage
drop of 5V seem to be economically acceptable. This
can be practically achieved as the test results in Figure
15 show. Here, the current utilization obtained in various tests is plotted versus the degree of regeneration
of the ion-exchange resin. Although the graphs show a
rather high degree of scatter, they do indicate that,
in general, the current utilization decreases with
increasing degrees of regeneration of the resin and that,
in the case of lead as well as copper recovery, current
utilization of more than 50% can be achieved at a
regeneration rate of about 50%.

CONCLUSIONS

The electrolytic regeneration of a Pb^{++}- or Cu^{++}-
ion charged cation-exchange resin seems to be technically
and economically feasible, when the regeneration is stopped before a breakthrough of H^+-ions occur. That means
that the regeneration process has to be stopped before
the regeneration is completed. In practice, a 50%
regeneration of a Pb^{++}- or Cu^{++}-ion charged resin can
be achieved at a current utilization of more than 50%.

REFERENCES

1.   Spiegler, K. S., in "Ion-Exchange Technology," Eds.:
     Nachod, F. C., Schubert, J., Academic Press, New
     York (1956).
2.   Pearson, R. G., US-Patent 2812300 (1957).

AN ULTRAFILTRATION DEVICE FOR FREE/BOUND DRUG ANALYSIS

Charles W. Desaulniers[1], Shmuel Sternberg, Ph.D.[1]
C.E. Pippenger, Ph.D.[2], and C.M. Garlock[2]

[1]Millipore Corporation, Bedford, Massachusetts
[2]College of Physicians and Surgeions, Columbia University
New York, New York

INTRODUCTION

A new ultrafiltration device, ULTRAFREE[R], has been developed
for the separation of analytical volumes of plasma water from human
plasma. While the experimental work described in the present paper
is restricted to five antiepileptic drugs, the device appears to be
useable for a wider range of drugs.

It has long been recognized that protein bound drugs are not
capable of producing a therapeutic response since only unbound or
free drug is in equilibrium across cell membranes. The bound
fraction serves as a reservoir for maintaining free drug at therapeutic concentrations. The free drug concentration, however, may
vary, depending on the total concentration, the concentration of
binding protein, competitive binding in the presence of other drugs,
disease states, and variations in individual metabolism.

Most work characterizing free/bound ratios or free drug
concentrations has been done by equilibrium dialysis, a good method,
but one which is time consuming, requires the measurement of the
concentration in both compartments of the apparatus and a mathemati-
cal treatment of the measured values to derive the original free
drug concentration in the plasma. The method thus is not practical
for routine analysis in the clinical laboratory where speed and
cost are significant factors.

The current clinical approach is to measure total drug
concentration (free and bound) and to either infer free drug
concentration from published data on binding, or to simply use the
total drug concentration as a therapeutic guide. Neither method is

159

as satisfactory as a direct measurement of the free concentration.

The reliability of a free drug estimate is questionable since there is wide individual variability in binding capacity. For instance, in the present study, free phenytoin concentrations ranged from 5% to 35% of the total.

Ultrafiltration has promised for many years to be an improvement in terms of time. However, cumbersome and expensive devices using pressure cells, gas sources, magnetic stirrers and expensive membranes have proved to be a barrier to widespread acceptance of the technique. Also, the high g forces in centrifugal devices make leakage a problem.

Various studies have also indicated, in the past, differing results between ultrafiltration techniques and equilibrium dialysis. It is important to realize that a 5% passage of protein during ultrafiltration of a drug that is 90% bound will lead to results which are 40% higher than the true value. Thus membranes or devices which are 95% retentive for albumin, for instance, are not good enough for separation of highly bound species. The more highly bound the drug, the more sensitive we are to leakage. This may explain why free/bound Calcium determination has been one of the more successful achievements of ultrafiltration; about 50% of the Calcium is bound and a 5-10% leakage is tolerable. The present device ultrafilters in an unstirred or dead-ended mode, and is shown in Figure 1.

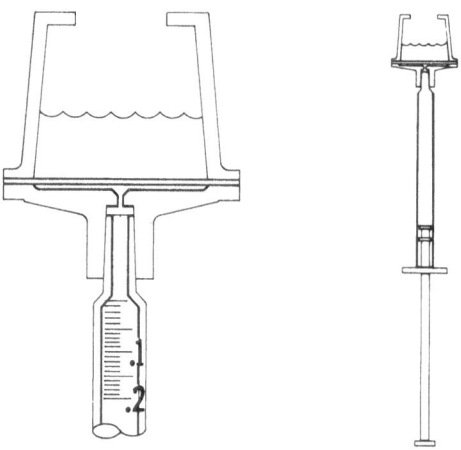

Figure 1.   Schematic of ULTRAFREE^R device.

EXPERIMENTAL

A 1 ml (1000 µl) disposable tuberculin syringe (BD 5602 plastipak or equivalent) is affixed to the female luer fitting on the bottom of the device. A 1-2 ml plasma sample is added to the cup, the syringe withdrawn to the 1 ml position, locked in place and the filtration proceeds.

Dead-ended ultrafiltration can be quite useful. Devices exist for concentration of protein solutions such as the Millipore CX-10 and CX-30 immersibles and the Amicon Minicon[R] and Centriflo[R] devices, but only the Centriflo[R] permits collection of ultrafiltrate, and its protein passage is a serious deficiency where a clean separation is desired.

The ULTRAFREE[R] unstirred disposable device containing an ultrafiltration membrane of about 40,000 M.W. cut-off delivers 250 µl of a 99.5% albumin free ultrafiltrate in 35-40 minutes, which may then be used for drug analysis.

Data are presented from clinical trials involving 206 plasma specimens obtained from patients receiving anti-epileptic drugs for seizure control. Ultrafiltration and equilibrium dialysis are compared as methods for isolating free drug, and, by subsequent analysis, determining plasma free drug concentrations.

Anti-epileptic drug concentrations in biological fluids may be measured by GLC, HPLC, RIA, spectrophotometry-fluorometry or the SYVA EMIT[R] technique. All analyses in this study were done using the EMIT-aed[R] homogeneous immunoassay procedure, modified so as to permit measurement of the very low drug concentrations found in the ultrafiltrate. For example, 50 µl of an ultrafiltrate at 0.3 ug/ml concentration will contain only 15 nanograms of the drug. The modification was a simple 10-fold calibration dilution where appropriate. The phenytoin curve, for instance, covered the range 0.25 ug/ml to 3.0 ug/ml instead of the usual 2.5 to 3.0 ug/ml. The method was originally developed for total drug assay and the usual calibration standards were of too high a range for some of the free drug determinations. No loss of precision or sensitivity was observed as a result of this modification.

Equilibrium dialysis experiments were carried out in 1 ml half cells rocked for 16 hours at room temperature. The membrane was a 10,000 mw regenerated cellulose (Technilab Inst.) and the buffer was a 0.067 M Sorensen's Phosphate buffer, pH = 7.4, with NaCl added to physiologic osmolarity (280-300 osm.). Drug concentrations were measured on the plasma and buffer sides at equilibrium.

Ultrafiltrations were done on a second aliquot using the ULTRAFREE[R] device. The free drug in the ultrafiltrate was

measured. Total drug, albumin, and total protein concentrations
were measured on a third aliquot of the original plasma.

All analyses were done in duplicate and a third replicate was
done where the analyses differed by more than 10%. The plotted
values are the averages of these replicates.

Plasmas were generally run within one or two days of collection.
A separate study of pooled plasmas over five days time indicates no
significant drug loss or change in free drug concentration.

The concentration of free drug in the ultrafiltrate is measured
directly. For the equilibrium dialysis experiment Figure 2
demonstrates how the original free drug, FD_O, is calculated from the
measurement, at equilibrium, of the total drug concentration, TD_{PS},
on the plasma side of the membrane and the free drug concentration,
FD_{DS}, on the dialysis side of the membrane.

Phenobarbital was used for clarity of explanation since it is
about 50% bound. It is this derived free drug concentration, FD_O,
that has been correlated with the directly measured ultrafiltrate
free drug concentration. Note that at equilibrium, the free drug

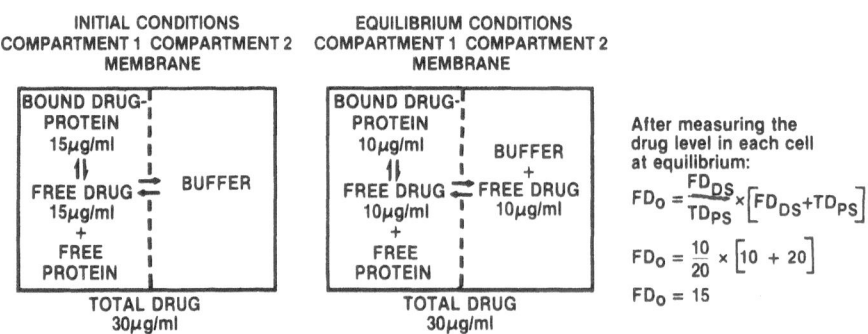

Figure 2.   Equilibrium dialysis of phenobarbital: model calculations
             of original plasma concentration from equilibrium conditions.
             conditions.

concentration is the same in both compartments and that the ratio
of free to total drug in the plasma side is unchanged from that of
the original plasma.

Figure 3 shows the correlation of the free drug concentrations
using the two methods for phenytoin, the most common antiepileptic
drug._ While there is some scatter, (the coefficient of variation,
S.D./$\bar{x}$, for EMIT-aed[R] is reported to be 10%), the slope 0.999 is
very close to 1 and the intercept is essentially 0. The correlation
coefficient is 0.954 and the standard error is 0.303.

Similar curves are shown in Figures 4, 5, and 6 for
phenobarbital, carbamazepine and primidone. There is somewhat more
scatter for carbamazepine with a correlation coefficient of 0.942,
a slope of 1.10 and an intercept of 0.224. Although equilibrium
dialysis was used as a reference method, the data infer nothing of
the accuracy or precision of one method over the other.

The standard errors are not directly comparable between drugs,
being a function of the magnitude of the quantity being measured.

A fifth drug was studied, ethosuximide, but is not reported
here since it is not protein bound. Thus ultrafiltration was not
necessary to determine the free concentration.

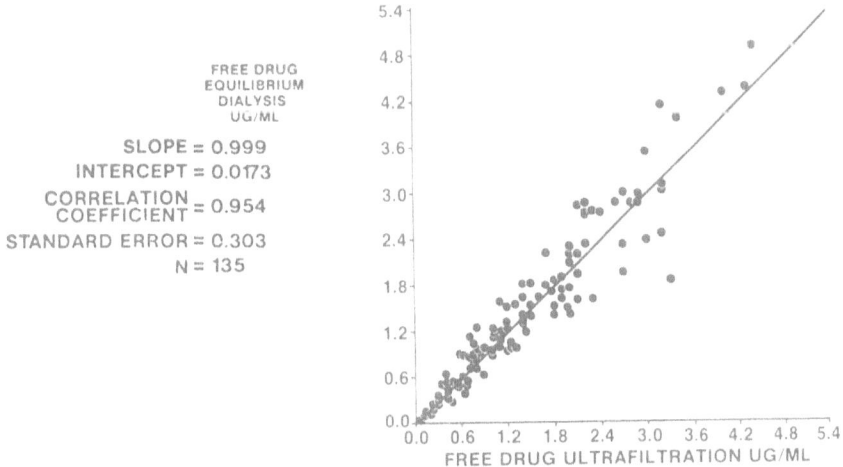

Figure 3.   Free pheytoin: ultrafiltration versus equilibrium
            dialysis.

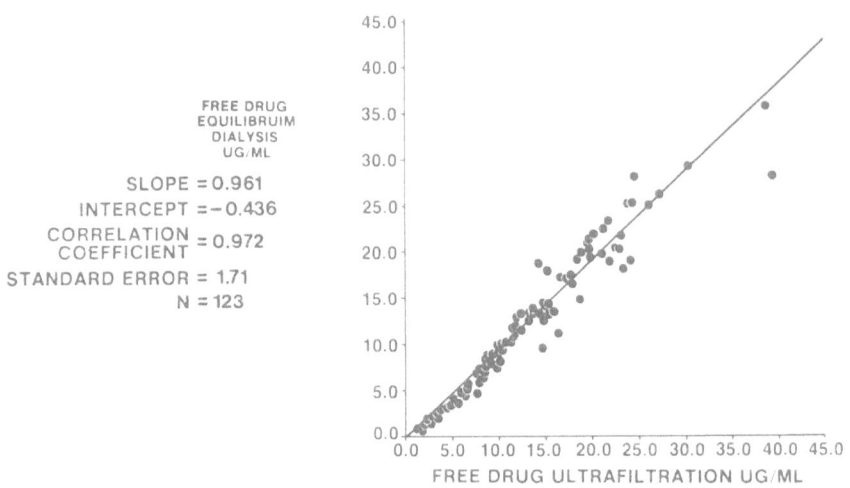

Figure 4.   Free phenobarbital: ultrafiltration versus equilibrium
             dialysis.

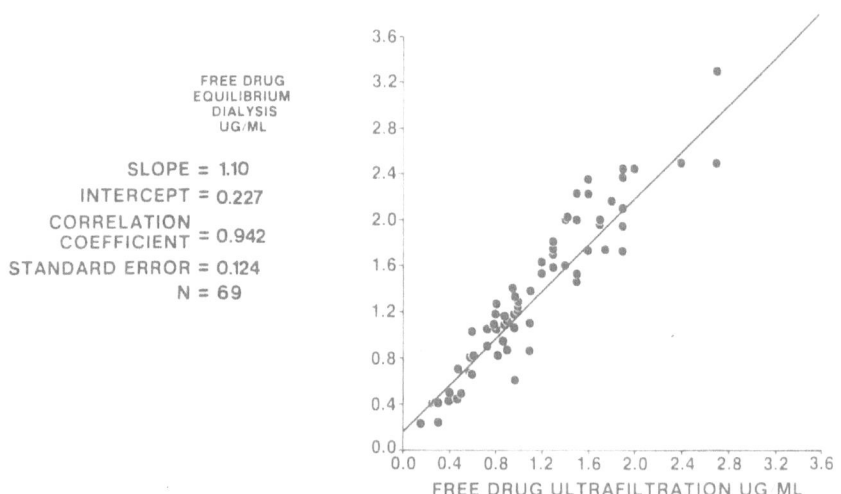

Figure 5.   Free carbamazepine: ultrafiltration versus equilibrium
             dialysis.

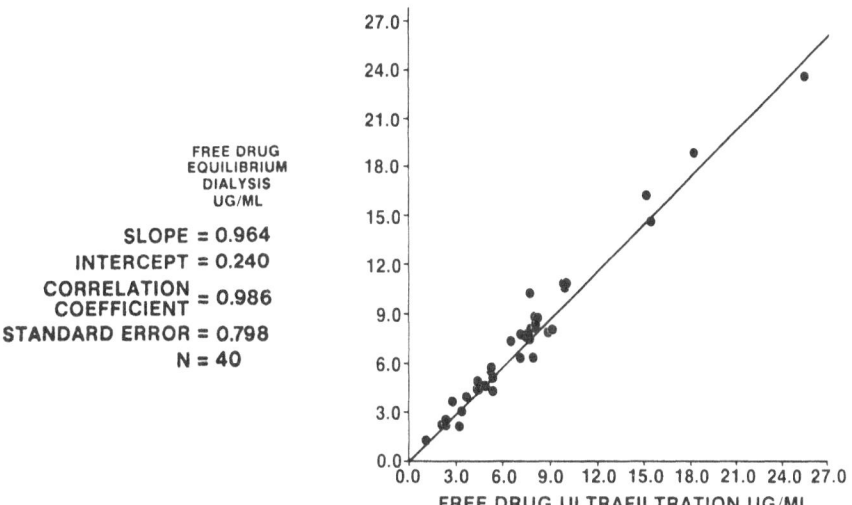

Figure 6.   Free primidone: ultrafiltration versus equilibrium
            dialysis.

FREE BOUND EQUILIBRIUM

Early in the work, there was concern about the effect of
equilibrium dialysis and ultrafiltration on binding, and on the
nature of the binding.

In other studies, the equilibrium between purified albumin and
drugs has been determined and dissociation constants determined.
In the present work, the patient plasmas are used to derive "in
vivo" equilibrium constants, by measuring total drug, free drug,
free albumin and bound albumin. A computer statistical program
developed at Millipore by C. Gifford was used to carry out what
would have been tedious calculations.

It is interesting to note that in deriving a typical equilibrium constant for a phenytoin plasma

$$K_{DISS} = \frac{(Unbound\ Drug)\ (Unbound\ Albumin)}{(Bound\ Drug\ Albumin)}$$

$$K_{DISS} = \frac{(7.1 \times 10^{-6}\ \frac{moles}{liter})\ (6.7 \times 10^{-4}\ \frac{moles}{liter})}{(7.2 \times 10^{-5}\ \frac{moles}{liter})} = 6.6 \times 10^{-5}\ \frac{moles}{liter}$$

that approximately 10% of the drug is free. Further 90% of the drug
is bound to 10% of the albumin, i.e., 90% of the albumin is free.
The ratio of free albumin molecules to free drug molecules is about
100/1. Thus, most of the phenytoin is bound (0.9 mole fraction
bound) and most of the albumin is unbound (0.1 mole fraction bound).

The mean dissociation constant for phenytoin was 6.7 ± 2.4
X 10^{-5} moles/liter for 135 individuals. The total phenytoin levels
ranged from 1.0 ug/ml to 55 ug/ml. The free phenytoin ranged from
0.03 ug/ml to 4.9 ug/ml. The albumin concentration ranged from
3.5 g/dl to 6.4 g/dl. In spite of the wide range of drug concentrations (5000%) and albumin concentrations (180%) the equilibrium
constants calculated are remarkably consistent, and the constants
derived are a good measure of the association of albumin and drug
in human plasma. Similar calculations for each of the other drugs
gave K_{DISS} for phenobarbital = 6.8 ± 2.4 X 10^{-4} moles/liter, K_{DISS}
for carbamazepine = 2.5 ± 1.2 X 10^{-4} moles/liter, K_{DISS} for
primidone = 2.8 ± 1.3 X 10^{-3} moles/liter.

The relative magnitude of the constants parallels the reported
binding, e.g., phenytoin ca. 90% bound, carbamazepine ca. 80% bound,
phenobarbital ca. 50% bound, and primidone ca. 20% bound.

K_{DISS} was also calculated using ultrafiltration data only. The
comparison of ultrafiltration and equilibrium dialysis derived data
is shown in Figure 7.

DRUG	$K_{DISS.}$ U.F. MOLES/LITER	$K_{DISS.}$ E.D. MOLES/LITER	RELATIVE MAGNITUDE
PHENYTOIN	6.1×10^{-5}	6.7×10^{-5}	1
CARBAMAZEPINE	1.7×10^{-4}	2.5×10^{-4}	3.7
PHENOBARBITAL	8.3×10^{-4}	6.8×10^{-4}	10.1
PRIMIDONE	2.7×10^{-3}	2.8×10^{-3}	41.8

Figure 7. Equilibrium constants from U.F. and eq. dial. anticonvulsant drugs.

There is good agreement between the two sets of dissociation
constants indicating that ultrafiltration is a good method for
determining equilibrium constants. There was somewhat less scatter
in the ultrafiltration derived constants.

Since the filtration is dead-ended or unstirred, the problem
of the polarizing layer and its effect on the filtration rate and
ultrafiltrate free drug concentration must be considered.

The filtration rate is essentially constant after the first
four minutes (see Figure 8) and is consistent with a protein layer
15-25 molecules in depth. The steady state ultrafiltration rate is
about 5 μl/minute.

Inasmuch as the free drug in the plasma water must pass through
this gel layer with its many unassociated albumin molecules, there
was some concern that the free drug would be adsorbed within it.
This does not happen. Two possible explanations are suggested to
support the observation that there is no detectable loss of free
drug in ultrafiltration.

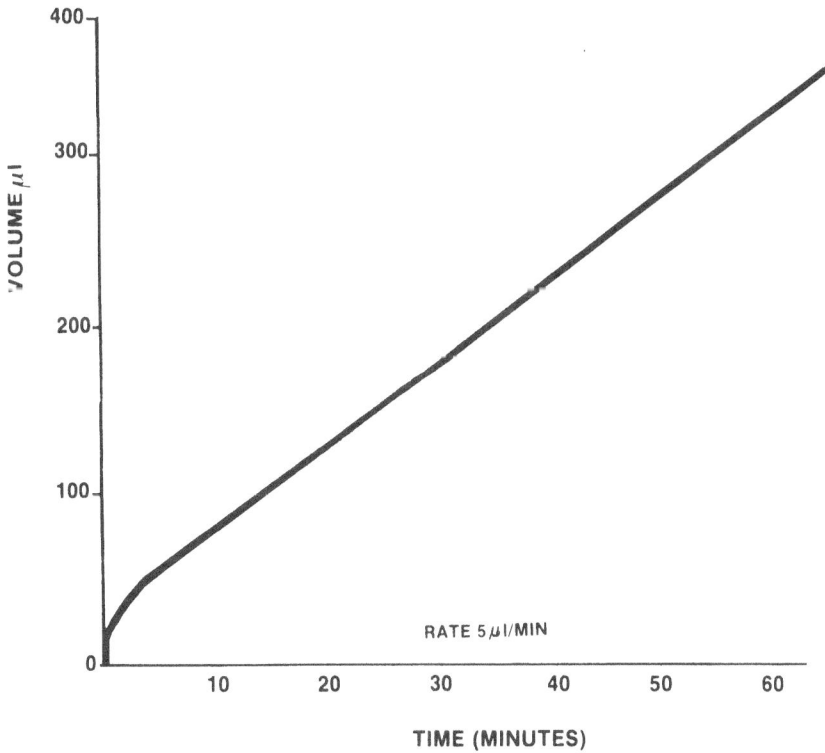

Figure 8. Ultrafree filtration rate of human plasma.

Since $K_{DISS} = \dfrac{(\text{Unbound Albumin})(\text{Unbound Drug})}{(\text{Bound Albumin \& Drug})}$

Let us represent the gel layer as being about 30% (the actual concentration is not necessary for this argument) protein or about four times that of the plasma.

Then $K_{DISS} = \dfrac{(4X\ \text{Unbound Albumin})(\text{Unbound Drug})}{(4X\ \text{Bound Albumin \& Drug})}$

Note that the gel layer contains bound and unbound albumin in the same ratio as is found in the plasma, thus the free drug concentration stays constant.

The other possible explanation is that, at these low ultra-filtration rates and with a relatively thin polarizing layer, any extra binding sites in the gel layer are quickly saturated and the reduction in free drug concentration in the plasma water is negligible.

REFERENCES

Ehrnebo, M., Agurell, S., Jalling, B. and Boreus, L.O. (1971)
Age difference in drug binding by plasma proteins. Studies on
human fetuses, neonates and adults.
Eur. J. Clin. Pharmacol 3: 189-193.

Odar-Cederlof, I. and Borga, O. (1976)
Impaired plasma protein binding of phenytoin in uremia and
displacement effect of salicylic acid.
Clin. Pharmacol. Ther. 20: 36-47.

Troupin, A.S. and Friel, P. (1975)
Anticonvulsant level in Saliva, serum and cerebrospinal fluid.
Epilepsia 16, 223-227.

H. Kutt (1978)
Evaluation of Unusual Antiepileptic Drug Concentrations.
In: Antiepileptic Drugs: Quantitative Analysis and Interpretation.
Edited by C.E. Pippenger, J.K. Perry and H. Kutt
p. 312-313 Raven Press, New York.

Booker, H.E. and Darcey, B. (1973)
Serum concentrations of free diphenylhydantion and their relation-
ships to clinical intoxication.
Epilepsia 14: 171-184.

Bochner, F., Hooper, W.D., Southerland, J.M., Eadie, M.J.
and Tyrer, J.H. (1974)
Diphenylhydantoin concentration in saliva.
Arch. Neurol. 31: 57-59.

Hooper, W.D., Bochner, F., Eadie, M.J. and Tyrer, J.H. (1974)
Plasma protein binding of diphenylhydantoin.
Clin. Pharmacol, Thir. 15: 276-282.

Lund, L., Berlin, A. and Lunde, P.K.M. (1970)
Plasma protein binding of diphenylhydantoin in patients with
epilepsy. Agreement between the unbound fractions in plasma
and the concentration in the cerebrospinal fluid.
Clin. Pharmacol. Ther. 11: 846-855.

Porter, R.J. and Layzer, R.B. (1975)
Plasma albumin concentration and diphenylhydantoin binding in man.
Arch. Neurol. 32: 298-303.

Reidenberg, M.M., Odar-Cederlof, I., von Bohr, G., Borga, O.
and Sjoqvist, F. (1971)
Protein binding of diphenylhydantoin and desmethylimipramine
in plasma from patients with poor renal function.
N. Engl. J. Med., 285: 264-267.

H. Kutt (1978) Evaluation of Unusual Antiepileptic Drug
Concentrations. In: Antiepileptic Drugs: Quantitative Analysis
and Interpretation, Edited by C.E. Pippenger, J.K. Perry and
H. Kutt. Raven Press, New York.

Borgia, O., Azarnoff, D.L., Forshell, G.P. et. al.
Plasma protein binding of tricyclic antidepressants in man.
Biochem. Pharmacol. 18: 2135-2143. (1969).

Borondy, P., Dill, W.A., Chang, T., Buchanan, R.A. and
Glazko, A.J. (1973)
Effect of protein binding on the distribution of 5.5-
diphenylhydantoin between plasma and red cells.
Ann. N.Y. Acad. Sci. 226: 82-87.

Bochner, F. et al. (1974) Arch. Neurol. 31: 57-59.

Schmidt, D. and Kupferberg, H.J. (1975)
Diphenylhydantoin, phenobarbital and primidone in saliva, plasma
and cerebrospinal fluid.
Epilepsia 16: 223-227.

Tropin, A.S. and Friel, P. (1975)
Anticonvulsant level in saliva, serum and cerebrospinal fluid.
Epilepsia 16: 223-227.

Lund, et al. (1970). Clin. Pharmacol. Ther. 11: 846-855.

Booker, H.E. and Darcey, B. (1973)
Serum concentrations of free diphenylhydantoin and their

relationships to Clinical intoxication.
Epilepsia 14: 171-184.

Hooper, W.D., Bochner, F., Eadie, M.J. and Tyrer, J.H. (1974)
Plasma protein binding of diphenylhydantoin
Clin. Pharmacol Ther. 15: 276-282.

McAuliffe, J.J., Sherwin, A.L., Leppick, I.E., Fayle, S.A. and
Pippenger, C.E. (1977)
Salivary levels of anticonvulsants. A practical approach to drug
monitoring. Neurology (Minneap.). 27: 409-413.

Chang, T., Dill, W.A. and Glazko, A.J. (1972)
Ethosuximide: Absorption, distribution and excretion. In:
"Antiepileptic Drugs", edited by D.M. Woodbury; J.K. Penry and
R.P. Schmidt. Raven Press. New York. pp. 417-423.

Browne, T.R., Dreifuss, F.E.; Dyken, P.R., Goode, D.J.,
Perry, J.R., Porter, R.J., White, B.G., and White, P.T. (1975):
Ethosuximide in the treatment of absence seizures. Neurology
(Minneap.) 25: 515-524.

Buchanan, R.A. (1972). Ethosuximide: Toxicity in: Antiepileptic Drugs, edited (ibid)

McAuliffe, J.J. et. al. Neurology (Minneap.), 27: 409-413.

Koch-Weser, J. (1974). Bioavailability of drugs. N. Engl. J. Med.
291: 233-237. 503-506.

Koch-Weser, J. (1972). Serum drug concentrations as therapeutic
guides. N.J. Med. 287: 227-231.

Kutt, H. and Penry, J.K. (1974). Usefulness of blood levels of
antiepileptic drugs. Arch. Neurol. 31: 283-288.

Booker, H.E. (1978). Clinical use and Interpretation of serum
phenytoin levels. In "Antiepileptic Drugs". p. 253-260. Edited
by Pippenger, C.E., Penry, J.K. and Kutt, H. Raven Press -
New York.

Koch-Weser, J. and Sellers, E.M. (1976). Binding of drugs to
serum albumin. N. Engl. J. Med. 294: 311 & 526.

Booker, H.E. and Darcey, B. (1973). Serum concentrations of free
diphenylhydantoin and their relationship to clinical intoxication.
Epilepsia 14: 177-184.

Hooper, W.D., Dubetz, D.K., Bochner, F., Cotter, L.M., Smith, G.A., Eadie, M.J. and Tyrer, J.H. (1975) Plasma protein binding of carbamazepine. Clin. Pharmacol. Ther. 17: 433-440.

Kutt, H. and Penry, K. (1974): Usefulness of blood levels of antiepileptic drugs. ·Arch. Neurol. 31: 283-288.

Penry, J.K., Smith, L.D. and White, B.G. (1972). Blood level determinations of antiepileptic drugs. Clinical value and methods. NINDS Bibliography Series No. 2 DHEW Publication No. (NIH) 73-396. U.S. Government Printing Office, Washington, D.C.

Sherwin, A.L., Robb, J.P. and Zechter, M. (1973). Improved control of epilepsy by monitoring plasma ethosuximide. Arch. Neurol. 28: 178-181.

Pippenger, C.E., Ph.D., J.K. Penry, M.D., H. Kutt, M.D., (1978) Antiepileptic Drugs Quantitative Analysis and Interp. Raven Press, New York.

Sherwin, A., M.D., Ph.D., Harvey, C.D., M.Sc., Leppik, I.E.,M.D. and Gonda, A., M.D., (1976) Correlation Between Red Cell and Free Plasma Phenytoin Levels in Renal Disease.

Reiderberg, M.M. (1973) Drug-Protein Binding and Interpretation of Measurements of Plasma Concentrations of Drugs. Hospital Pharmacy, p. 22 ff, 8, 1973.

Spector, R.D., Korlsen, T., Lorengo, A.V., (1972) A Rapid Method for the Determination of Salicylate Binding by the Use of Ultrafilters. J. Pharm. Pharmacology, 24, 1972, p. 786 ff.

Behm, H.L., Wagner, J.G., (1979) Errors in Interpretation of Data from Equilibrium Dialysis Protein Binding Experiments Res. Commun. Chem. Pathol Pharmacol 26 1, 1979.

LeComte, M., Zini, R., D'Athis, P., Tillement, J.P., (1979) Phenytoin Binding to Human Albumin Eur. Journ. Drug Melab. Pharmaco peret. 4, 1, 1979.

Kurz, H., Trunk, H., Wertz, B., (1977) Evaluation of Methods to Determine Protein Binding of Drugs Arzneim. - Forsch. 27, 7, 1977.

Blatt, W.F., Robinson, S.M., Bixler, H.J., (1968)
Membrane Ultrafiltration - The Diafiltration Technique
Applied to Microsolute Exchange and Binding Phenomena
Analytical Biochemistry 26, 151-173 (1968).

MICROPOROUS MEMBRANES AND CROSSFLOW FILTRATION OF MACROMOLECULES

AND PARTICLES

P.D. Lindley,* W.P. Olson,* and M.R. Faith[†]

* Hyland Therapeutics Division, Travenol Laboratories
  Los Angeles, CA 90039 and
† Institute for Cancer and Blood Research
  Beverly Hills, CA 90211

INTRODUCTION

Filtration most often is defined as the act of passing a fluid
through a porous solid in order to remove particles. The porous
solid and its holder is the filter. Polymeric membrane filters
(PMF), rated at a poresize of 0.2μm (diameter), commonly are used
in the complete removal of bacteria and molds from liquids such as
therapeutic blood fractions.[1] Positive pressure, supplied by a
pump or a gas such as nitrogen, usually drives the process.[2] Almost invariably these systems are unidirectional or deadend, whereby
all of the liquid feed addressed to the filter passes through until
particles have fouled the pores; the fouling process is termed clogging. We consider a filter clogged when the flow rate is 20% of
the initial rate.

In serum, plasma, or plasma fractions, the particles that clog
0.2μm-rated PMF are dirt, bacteria/molds, fibrin clots, and lipoproteins.[3] The foulants occlude pores of the PMF and accumulate on
the upstream side of the filter to form a cake; for practical purposes, the particles become part of the filter. The particulate
cake increases the resistance of the filter to the flow of liquid
because the filter becomes thicker and the pore density declines.
This phenomenon is particle polarization.[4] It is readily cured by
the use of fibrous or depth prefilters.[2] However, there are occasions when one wishes to recover the particles[5,6] or to prolong the
PMF life, and this is best done by tangential (crossflow) filtration.[4,7] Crossflow is distinct from deadend filtration in that a
proportion of the feed sweeps foulants from the filter surface.
This proportion of the feed that does not pass through the filter
is the retentate and usually is recycled into the container from

173

which the feed is pumped (Fig. 1). Crossflow commonly is applied
to large ultrafiltration and to all reverse osmosis systems[8] but is
uncommon with microporous (0.01 to 10μm poresize diameter) PMF.

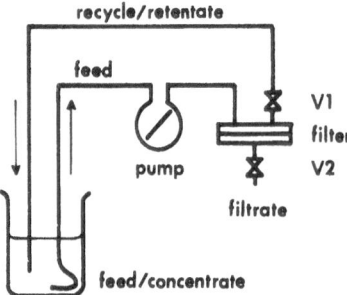

Fig. 1.   Twenty-five mm diam disc crossflow/deadend system. V1 and
          V2 are valves. For deadend operation V1 was closed and con-
          stant pressure was provided with N2.

        Macromolecules, such as proteins, tend to accumulate in a dense,
gel-like layer on the upstream side of ultrafiltration (UF) membranes
and, by virtue of the gel density and nonporous structure, it clogs
the membrane. This phenomenon is gel (or concentration) polarization
and crossflow is the most effective fluid management strategy.

        Dense, gel-like sheets of lactalbumin have been detected on
0.2 μm-rated track-etch polycarbonate filters (PC, a form of PMF from
Nuclepore Corp., Pleasanton, CA) when whey was filtered, deadend,
through the PC.[10] Somewhat similar sheets have been seen [11] when de-
mineralized water was filtered through PC. The amorphous materials
from the water likely were capsular polysaccharides sloughed from
bacteria entrained in ion exchange resins upstream of the PC. These

examples suggest that gel polarization also occurs with 0.2 μm-rated
PMF. Additional evidence is the frequent (but poorly documented)
observation that the filterability of serum or plasma through deadend
PMF is inversely related to pressure.

Viscosity of the feed limits the filterability of a liquid through
a deadend 0.2 μm rated PMF. Whether the liquid is a concentrated
solution of sucrose or of albumin, or is a motor oil, the relation
between viscosity and filtration rate through a PMF is given by

$$\frac{dV}{dt} = \frac{\pi r^4 \Delta P}{8\eta\ L}$$

when dV/dt is the flow rate at time t, r is the tube (pore) radius. η
is the fluid viscosity, and ΔP is the pressure drop over the PMF thickness L (Poiseuille's law).

We show in this chapter that crossflow filtration with 0.2 μm-
rated PMF alleviates the problems of viscosity-imposed limits to
flux, flow decay caused by protein adsorption, and pore occlusion by
particles.

MATERIALS AND METHODS

Filtration systems

The unit shown in Figure 1 consisted of a 25mm-diameter stainless
flowcell (catalogue no. 25700-01, Creative Scientific, Long Beach,
CA) driven with a peristaltic pump masterflex (model 7013, Cole-
Parmer Co., Chicago, IL). The filtration area in this cell was 4.15
cm^2 and the cell height was 39 mils (990 μm). During normal operation, cell pressures ranged from 5 to 12 psi. Valve V1 was open
during crossflow operation and closed for deadend filtration. Valve
V2 was sometimes shut during crossflow operation to facilitate the
recording of rate data. Filtration rates were estimated by timing
with a stopwatch the minutes required to collect 1 or 3 ml of filtrate into test tubes marked for volume. Recycle flow rates of the
retentate were estimated by timing the flow into a graduated cylinder.

On occasion a stainless 47 mm cell with a thick-walled Perspex
cylinder (catalogue no. 47920-00, Creative Scientific) was used for
deadend filtration studies; only N$_2$ was used to provide constant
pressure. The filtration area in the cell was 15.90 cm^2.

Membranes

Filters under test were mixed esters of cellulose (5 μm-rated,
SMWP 025, or 0.2 μm-rated, GSWP 025; Millipore Corp., Bedford, MA),
cellulose acetate (0.2 μm-rated, EGWP 025; Millipore), nylon 66
(0.2 μm-rated, NR02525; Pall Corp., Glen Cove, NY), or the PC men-

tioned earlier. A disc of Whatman 541 paper overlay each of the
membrane filters to prevent the filter from lifting into V1 during
crossflow operation. The 541 did have a significant protective
effect on filters studied in deadend mode, and may also have con-
tributed to undulating flow of the crossflow feed; it has been proposed that undulating flow contributes to the sweeping action during
crossflow operation.[12]

Solutions and Suspensions

Normal human serum albumin (20% w/v) was old and showed moderate
but significant polymerization. The polymerized albumin standard
(25% w/v) was very old material with a very high degree of polymerization. Originally, both had been filtered to 0.2 μm, or the equivalent, when they were produced. Antihemophilic factor was freeze-dried
material that had been prepared as a partially-purified cryoprecipi-
tate, then membrane-filtered before lyophilization. Yeast cells were
Fleischmann's Active Dry Yeast which were weighed and taken to the
appropriate volume phosphate-buffered saline. The various feeds and
filtrates were held briefly at +5°C until they were analyzed. The
exceptions were the AHF samples which were maintained at 20 to 25°C
since AHF tends to aggregate in the cold.

Analytical Methods

Proteins were estimated at 260 and 280 nm by the Warburg-Christian
method.[13] Electrophoresis for analysis of albumin polymers was in a
10% polyacrygamide slab in the Multiphor™ system (LKB Instruments,
Rockville, MD).[14] The tracking dye was bromthymol blue. Current
(about 190 mA and a field strength of about 6 V/cm) was stopped and
fixing commenced when the dye had migrated 6 cm from the origin.
Antihemophilic factor (AHF or clotting factor VIII) samples were
electrophoresed in slabs of 1% agarose in the Multiphor.[15] Staining
of the gels was in the recommended manner[14,15] with Coomassie Blue
G250. Gel filtration of albumin preparations by upward flow through
a 26 mm x 90 cm column of Sephacryl S-300SF (Pharmacia Fine Chemicals,
Piscataway, N.J.) was maintained at about 1.36 ml/min with a P-3 pump
(Pharmacia) and column effluent was monitored at 280 mm with a Uvi-
cord ultraviolet monitor (LKB) interfaced with a flatbed recorder.
Viscosity in centipoise was estimated with Cannon-Fenske kinematic
viscometers.[16] Approximation of the passage of yeast cells through
5 μm-poresize cellulose acetate and mixed esters of cellulose PMF was
with the HIAC particle analyzer (model no. PC-320, HIAC Instruments
Div., Pacific Scientific, Montclair, CA) with the 5 to 10 μm channel.

RESULTS AND DISCUSSION

Clean Albumin

When a moderately polymerized 20% albumin was filtered, deadened

or crossflow through the 25 mm crossflow cell, there was no detectable rate decay in crossflow but the deadend rate at 5 psi ceased
almost immediately (Fig. 2).

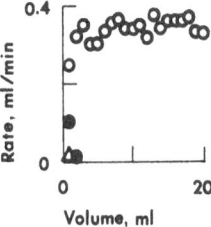

Fig. 2.   Filtration of moderately polymerized 20% albumin through
           0.2 μm poresize membranes. (Δ) Deadend filtration through
           Nylon 66; (●) deadend filtration through cellulose acetate;
           (o) crossflow filtration through cellulose acetate.

        An immediate question was, what effect did filtration have on
the polymer composition? The Sephacryl S-300SF protein profiles of
the original material, 5 psi deadend filtrate, and crossflow filtrate
and retentate were indistinguishable (Fig. 3).   From slab gel electro-
pheresis of the sample at 20g% one could detect trimer or tetramer;
however, these albumin polymers were present in deadend and crossflow filtrate and retentates (Fig. 4.)   There were viscosity differences between the original 20% product, the PMF crossflow filtrate
and crossflow retentate (Table 1).

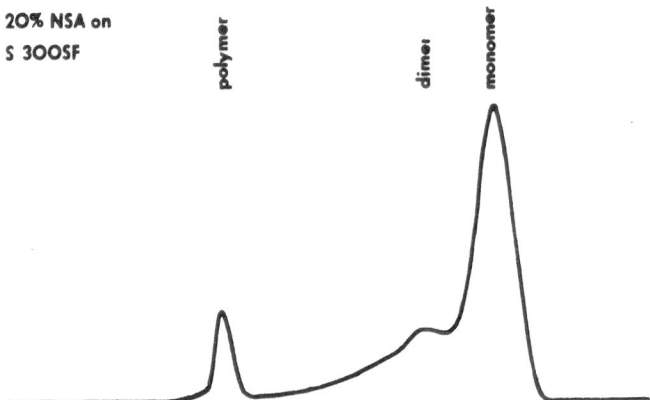

Fig. 3.   Sephacryl 300 superfine gel profile of moderately polymer-
           ized 20% albumin. The material was chromatographed as
           described in Analytical Methods.

     The differences were modest but consistent. When we later con-
structed a standard curve of g% v. viscosity in cP for nomomeric
albumin (Fig. 5) it became clear that the removal of very small
amounts of a polymer could reduce significantly the viscosity of the
product; it is to this subtle effect that we attribute our Table I
results. We also noted that the retentate was a golden brown, the
original material intermediate in color, and the filtrate somewhat
greenish. Polymerizing albumin becomes increasingly brown.   The
brown-to-green color shift often occurs with increasing pH but that
was not the case here.

     It now seemed prudent to test the conventional wisdom that dead-
end filtration of clean albumin at elevated pressures would correlate
with decreased throughput. It was not so for albumin. Rather,
throughput and rate increased with pressure (Fig. 6). As a first
approximation to mechanism we plotted reciprocal rate v. volume; a
straight line implies pore blocking at the filter surface.[17] We
also plotted reciprocal rate v. cumulative filtration time (not
shown), a straight line implies pore occlusion by accumulation of
materials within the pores, an event which slowly closes the pore as
though it were an irreversible iris diaphragm.[18] The reciprocal rate-

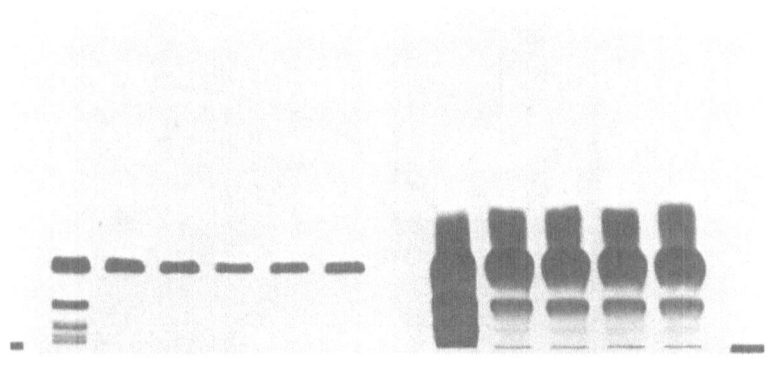

A B C D D E   A B C D E

Fig. 4   Polyacrylamide gel electrophoresis profiles of moderately
          polymerized 20% albumin. The samples on the left were run
          at 0.2 gram %; those on the right were run at 20 gram %.
          A, highly polymerized 20% albumin standard; B, unfiltered
          albumin; C, deadend filtrate; D, crossflow filtrate; E,
          crossflow retentate. For other details, see Analytical
          Methods.

Table I. Viscosity of Albumin Solution

Solution	Treatment	Viscosity
20% Normal Serum Albumin	None	2.830
Polymerized 20% NSA	Unfiltered	7.405
" " "	Retentate	7.893
" " "	Filtrate	6.706

volume plots were essentially linear. Inspection of Fig. 6 reveals
that reciprocal flux as a function of cumulative time was essentially
linear at 28 psi and that result seems inconsistent with the clogging
of a 0.2 μm-diameter pore by macromolecules adsorbing to the filter.
It is possible that particles approaching 0.2 μm in diameter remained

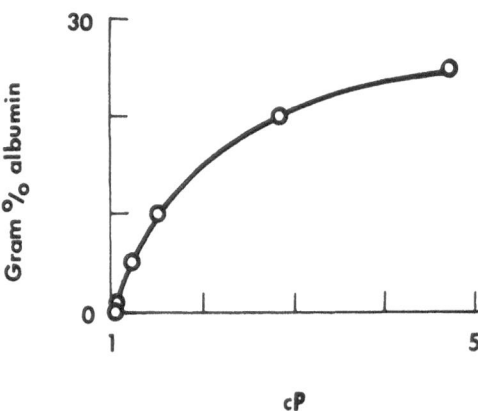

Fig. 5:   Albumin concentration in gram % vs. viscosity in centi-
          poise for monomeric albumin.

in the product after the original filtration to 0.2 μm; at high
pressure these particles followed the fluid streamlines into the
pores and clogging by particles (CBP) occurred. Alternatively,
microporous adsorptive clogging (MAC) and CBP may be indistinguishable for viscous fluids at high differential pressures.

       The MAC appears to be more dramatic at low pressures (=low flow
rates). Electrokinetic and hydrophobic effects, leading to the
immobilization of proteins on the filter, commence when protein and
surface are Å-distances apart. Thus, MAC must be facilitated as
dwell time of protein in the filter increases. In this regard MAC
is distinct from gel polarization which is facilitated by high pressures and rates.

Clean AHF

       We then examined the filtration of clean AHF through 0.2 μml-
poresize nylon PMF (Fig. 7). Analytical plots of deadend filtration
through the filter at 5 psi (Figs. 7 and 8) indicated MAC assays of
the feed and filtrates showed no significant change in AHF concentration, whether the material was filtered deadend, crossflow, or not
at all. Electrophoresis in 1% agarose (Fig. 9) showed no

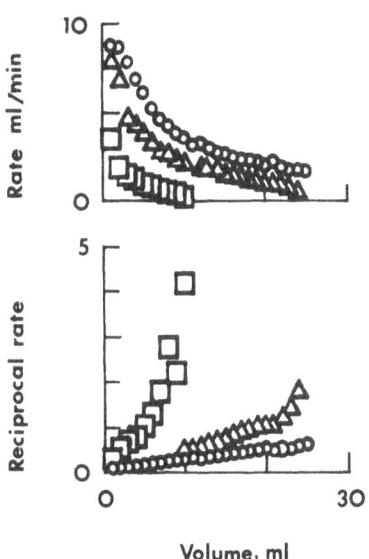

Fig. 6   Filtration of clean 20% albumin through 0.2μm poresize mixed
           esters of cellulose filters at different pressures (□) 10
           psi; (Δ) 20 psi; (0) 28 psi.

apparent change in the ratio of fibrinogen to AHF between feed and
filtrate, whether filtration was deadend or crossflow.

    These results were consistent with the AHF crossflow-deadend
filtration studies of Stampe,[19] who found that AHF crossflow increased
throughputs, and titer was not lost. The results also implied the
accumulation of a layer of protein on or in the filter, leading to
slow occlusion of the pores. Scanning electron microscopy (SEM) of PC
through which AHF had been deadend-filtered revealed protein deposits
(Fig. 10) which became confluent over the filter as clogging prog-
ressed (Fig. 11). Elemental analysis by energy dispersive x- ray
analysis indicated the presence of S, the only source of which
was the methione and cysteine of the plasma proteins. The SEM
results were similar to those obtained with high-purity water

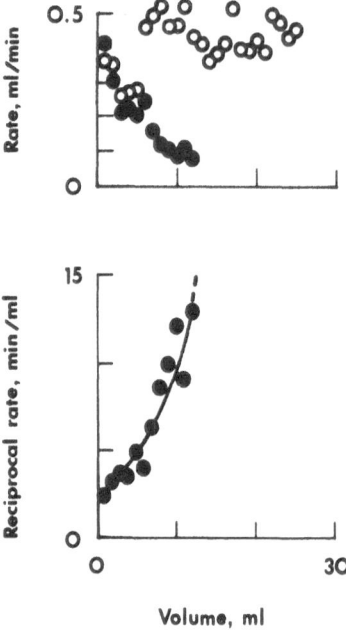

Fig. 7   Filtration of clean AHF through 0.2 μm poresize Nylon PMF.
         (•) Deadend filtration; (o) crossflow filtration.

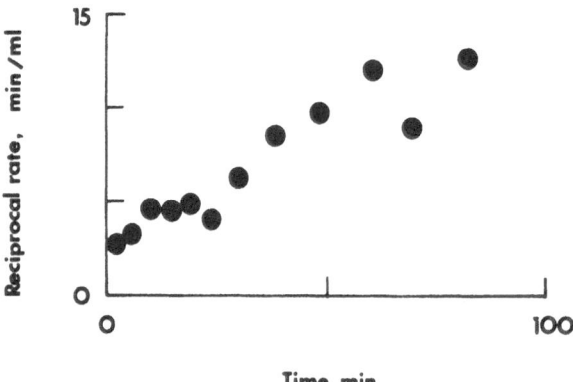

Fig. 8   Reciprocal rate vs. time plot for clean AHF filtered deadend
         through 0.2 μm poresize Nylon PMF.

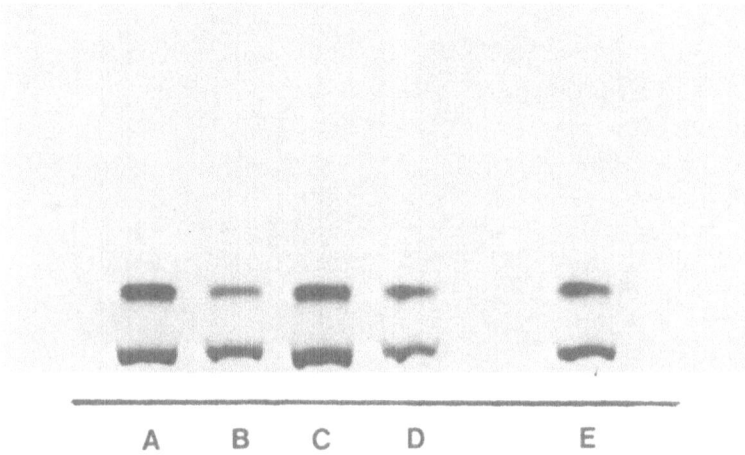

Fig. 9   Ararose gel electrophoresis profiles of AHF. A, unfiltered
         AHF used in crossflow filtration studies; B, corssflow fil-
         trate; C, corssflow retentate; D, unfiltered AHF used in
         deadend filtration studies; E, deadend filtrate.

containing capsular bacterial polysaccharides[11] and lead us to specu-
late that large macromolecules, like fibrinogen, tend to follow the
fluid streamlines and many of the macromolecules impact into the filter surface immediately around the pores to form ring or doughnut-like
deposits (Fig. 12). A model for MAC, perpendicular to the filter
surface, is identical with that of Blatt et al.[9] for gel polarization
on UF.

If the model is accurate, then crossflow cures the initiation of
MAC by interfering with doughnut formation. However, MAC still may
occur below the surface of a PMF, in which case long-term crossflow
filtration of clean albumin or AHF will show flow decay; we expect
that this will occur.

The literature indicated that: (A) protein binding to a surface
including membranes is hydrophilic and reversible[20]; (B) hydrophobic
surfaces bind far more protein than hydrophilic surfaces[21] (even
Teflon eventually is covered to 90% by fibrinogen or 20% by albumin[22]); (C) although albumin and fibrinogen compete for membrane binding sites[21], fibrinogen[23] or hemoglobin[24], when present, likely come
to predominate on the surface; and (D) the total number of binding
sites for various proteins, per unit area on a given surface likely
is constant[25].

X14000   Filter #8

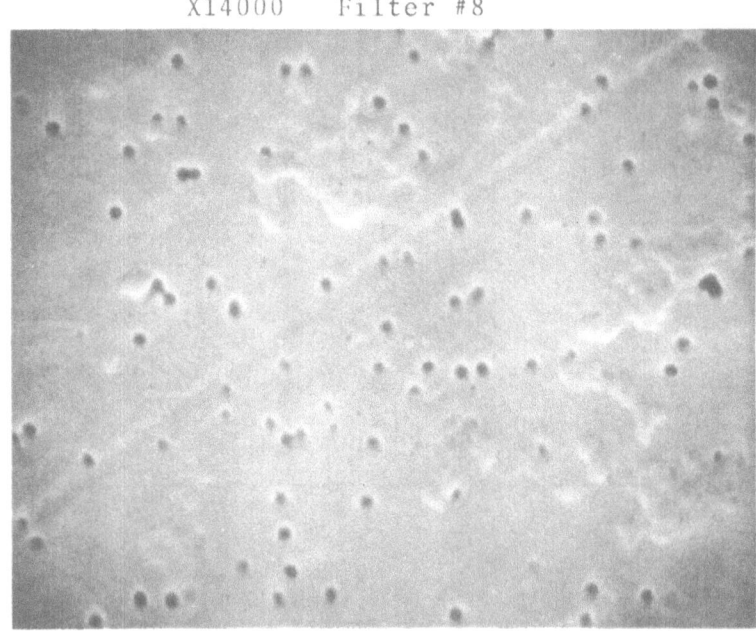

Fig. 10 Scanning electron micrograph of a 0.2 μm poresize poly-
         carbonate filter through which AHF had been deadend filter

X14000   Filter #9

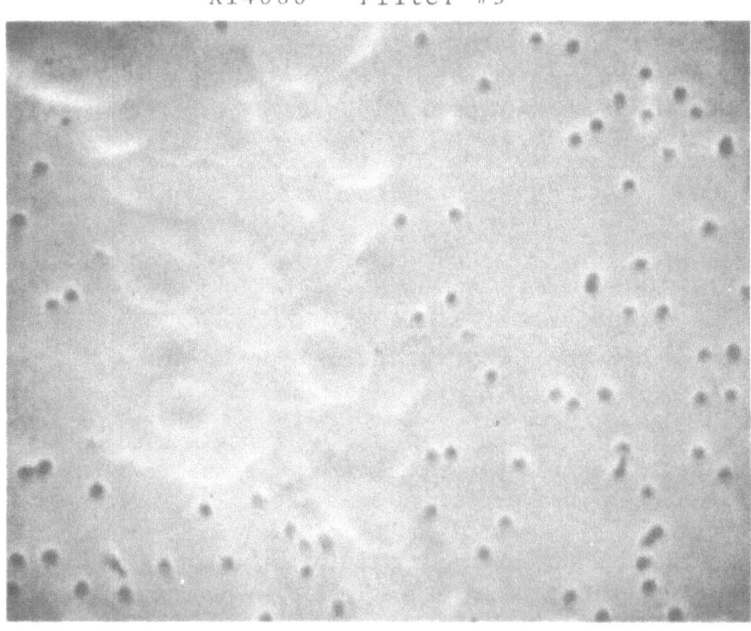

Fig. 11 Scanning electron micrograph of a 0.2 μm poresize polycar-
         bonate filter through which AHF had been deadend filtered.

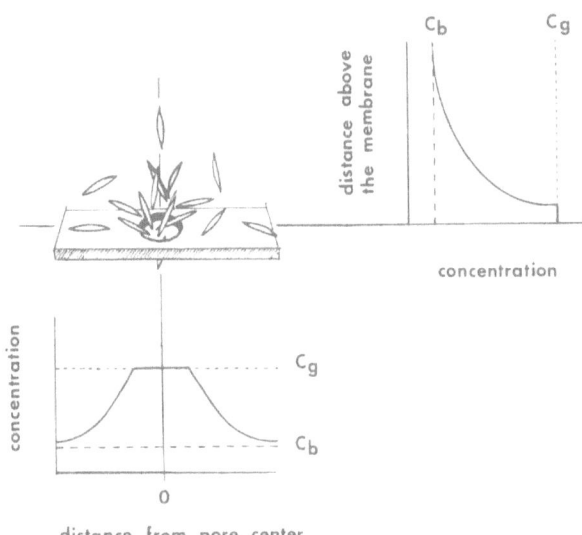

Fig. 12   Illustration of the model for microporous adsorptive clog-
          ging (MAC).   See text for explanation.

From the literature one might conclude that deadend filtration of
complex protein mixtures must result in significant changes in composition because MAC occurs.   At least for AHF this appears not to be
so.   On the other hand, the macromolecular composition of serum or
plasma, filtered in deadend mode through 0.1 or 0.2 µm PMF, does
change. [3,26]   Even track-etch polycarbonate (PC), which binds far less
protein than other filters,[27] clogs in crossflow.[3]   The practical conclusion is that one ought not to perform 0.2 µm-poresize filtration of
a complex protein suspension in line with a filling machine until it
is established that MAC does not cause a major variance in filtrate
composition during process.

Shear reduces significantly the viscosity of non-newtonian fluids
such as whole blood, plasma, serum or fractions thereof.[28]   Consequently, the shear generated during crossflow reduces the viscosity of
the feed and the filtration rate increases.   The same mechanism likely
obtains for aggregated particles which dissociate when exposed to high
shear, e.g., viruses and thixotropes such as finely divided amorphous
silica.

Deformable Particles

Rigid particles, like diatomaceous earth (DE), when suspended in
clean water, produce a highly porous cake on a PMF and add little to
membrane resistance to flow.   Crossflow rates are uniform and unevent-
ful.   The role of DE in the deadend filtration of particles from

protein-containing suspensions such as beer, has been described in considerable detail.[29]

Deformable particles, such as large flocs of protein and yeast cells, rapidly clog PMF at the filter surface. Our data on yeast capture by PMF confirm the linearity of plots of reciprocal rate v. cumulative volume. The crossflow plots are bimodal which indicates that at some point in the filtration there is a marked decline in the rate at which the filter clogs. This is true in the crossflow fil- trative concentration yeast over 0.2 µm and 5 µm-poresize PMF. The only difference between the curves is that the 0.2 µm PMF clogs far more rapidly than the 5 µm membrane. However, if the original curves or the reciprocal rate v. volume plots are normalized, they can be superimposed.

The yeast cells do not pass through a 0.2 µm-poresize PMF but a proportion of the cells pass through a 5 µm membrane, even during crossflow (Fig. 13). When the initial feed to a 5 µm membrane con- tained about 950 yeast cells/ ml, about 300 cells/ml appeared in the first 15 ml of filtrate. Cells in filtrate declined significantly only at the transition from rapid to slow clog.

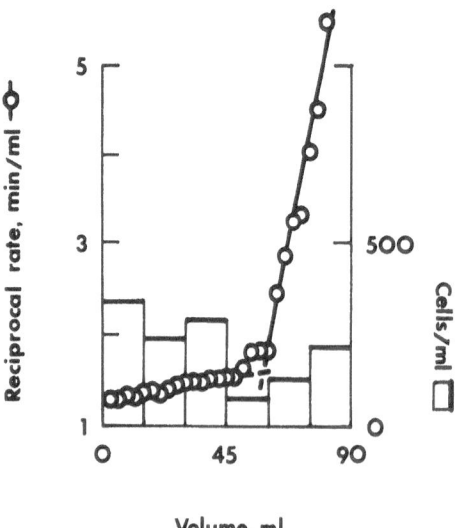

Fig. 13   Crossflow filtration of yeast cells through some 5 µm cellu- lose acetate membrane filter.

We interpret reciprocal rate results as indicating a change in the permeability of the filter and the caked yeast. The initial rapid clogging shown in Fig. 13 may occur as yeast cells are immobilized in a substantial proportion of the larger pores.

Because the bulk of liquid flow is through the largest pores (which offer the least resistance to flow), complete occlusion of those pores is consistent with a rapid decline in flow rate. Additional cells may block small pores or form a second layer upon the first; in either case, the effect on flow rate must be less dramatic.

The passage of 5 μm-diameter cells through a 5 μm-poresize PMF, operated in crossflow mode, dashed a hope that the crossflow strategy would cause PMF to retain all particles with a diameter equal to the membrane poresize.

Concentration of bacterial cells over 0.2 μm or over UF membranes also can be bimodal. Initially, the PMF rate is very high but it soon equilibrates to a rate _less_ than one obtains through an ultrafilter.[30, 31] It is possible that the microscopic roughness of the PMF surface contributes to the retention of organisms in the primary cake.

Our final thoughts on particle retention and concentration with PMF crossflow are that when the particles are to be retained on the filter and discarded, depth filtration upstream of the PMF usually is necessary and there may be no advantage to crossflow. However, if the particles, such as bacteria, are to be recovered, PMF or UF crossflow is a desirable strategy. For heat labile, viscous liquids, crossflow clearly is advantageous.

Crossflow Systems

The 25 mm crossflow disc system shown in Fig. 1 also is available from Creative Scientific in 90 and 142 mm diameters with unique sanitary quick connects. Millipore Corp., Bedford, MA and Sartorius GmbH, Gottingen, GFR, produce crossflow systems for UF or PMF sheetstock in parallel.

Cartridge crossflow systems are available from Gelman Sciences, Ann Arbor, MI in standard or disposable models.[32] The differences between deadend and crossflow cartridges are shown in Fig. 14 which is a sectional view. The essential elements for crossflow with pleated PMF cartridges are: an impermeable sleeve that covers the outside of the pleats save for the top and bottom which are left exposed; a collar that marries the sleeve to the housing; and a means of recycling the retentate that flowed through the pleats to emerge at the bottom of the sleeve. The system forces the feed to pass rapidly over the length of the cartridge but within the pleats so that shear is developed at the membrane surface.[33]

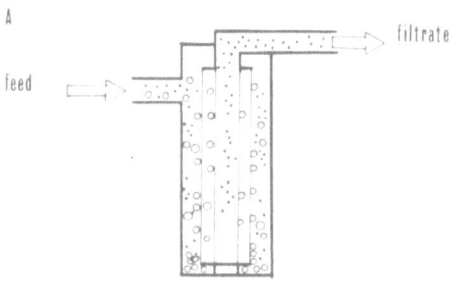

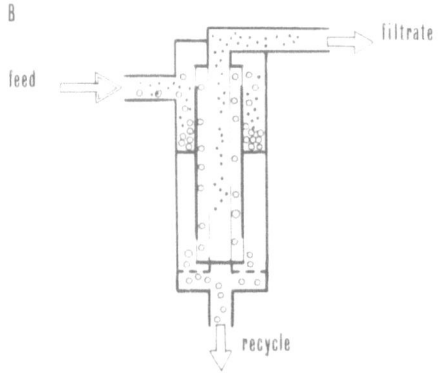

Fig. 14   Sectional view of deadend and crossflow cartridges.   See
          test for explanation.

ACKNOWLEDGMENTS

     We are most grateful to John Bright of LKB Instruments who
provided us with help and advice on the electrophoresis, as did Ray
Berkebile, Research Dept., Hyland Therapeutics (HT).   Charles
Barksdale, and Siony Choquette, Development Dept., HT, performed
column and AHF activity assays, respectively.   Dr. Ted Meltzer,
Gelman Sciences explained to us the merits of reciprocal rate plots.
Most of all, we are grateful to Steve Holst and Linda Sarno, HT, for
support of the project.

REFERENCES

1.   W. P. Olson, Pharm. Technol. 3(11):   84 (1979)
2.   J. V. Fiore, W. P. Olson, and S. L. Holst, Depth Filtration, in:
     "Methods in Plasma Fractionation," J. Curling, ed., Academic
     Press, Ltd., London (1980).
3.   W. P. Olson and M. R. Faith, Prep. Biochem. 8:   379 (1978)
4.   J. D. Henry, Cross Flow Filtration, in: "Recent Developments in
     Separation Science, Vol. II," N. N. Li, ed., Chemical Rubber Co.,

Cleveland, Ohio (1972)

5. R. C. Lukaszewics, A. Korin, and D.I. Hauk, Manuscript in preparation (1980). Data have been presented at the national meeting of the Parenteral Drug Assoc., New York Hilton, NYC, Nov. 3-5, 1980.

6. D. E. Reid and C. Adlam, J. Appl. Bacteriol 41: 321 (1974).

7. W. L. Bickmore, R. A. Champa, and W. P. Olson, Pharm. Technol. 1 (6): 55 (1977).

8. M. C. Porter, Membrane Filtration, in: "Handbook of Separation Techniques for Chemical Engineers,: P. A. Schweitzer, ed., McGraw-Hill, New York (1979).

9. W. F. Blatt, A. Dravid, A. S. Michaels, and L. Nelsen, Solute polarization and cake formation in membrane ultrafiltration: Causes, consequences and control techniques, in: "Membrane Science and Technology,: J. E. Flinn, ed., Plenum Press, New York (1970).

10. D. N. Lee and R. L. Merson, J. Food Sci. 41: 403 (1976).

11. W. P. Olson, Ind. Water Eng. 16(1): 20 (1979).

12. "1977 Millipore Ultrafiltration Membranes and Systems," PB 819, Millipore Corp., Bedford, MA (1977).

13. O. Warburg and W. Christian, Biochem. J. 310: 384 (1941-1942).

14. "SDS and Conventional Polyacrylamide Gel Electrophoresis with LKB 2117 Multiphor," Application note 306, LKB-Produkter, Bromma, Sweden (1977).

15. "Agarose Gel Electrophoresis with LKB 2117 Multiphor,: Application note 310, LKB-Produkter, Bromma, Sweden (1977).

16. M. R. Cannon and M. R. Fenske, Ind. Eng, Chem., Anal. Ed. 10: 680 (1939).

17. P.H. Heramans and H. L. Bredge, Rec. Trav. Chim. 54: 680 (1939).

18. B. D. Bowen, S. Levine, and N. Epstein, J. Colloid Interface Sci. 54: 375 (1976).

19. D. Stampe, Pharm. Ind. 38: 113 (1976).

20. W. J. Dillman and I. F. Miller, J. Colloid Interface Sci. 44: 221 (1973).

21. H. Y. K. Chuang, W. F. King, and R. G. Mason, J. Lab. Clin. Med. 92: 483 (1978).

22. R. C. Eberhart, L. D. Prokop, J. Wissenger, and M. A. Wilkov, Trans. Am. Soc. Artif. Int. Organs 23: 134 (1977).

23. B. W. Morrissey, Ann. N.Y. Acad. Sci. 283: 50-64 (1977).

24. T. A Horbett, P. K. Weathersby, and A.S. Hoffman, Thromb. Res. 12: 319 (1978).

25. L. A. Cantarero, J.E. Butler, and J. W. Osborne, Anal. Biochem. 105: 375 (1980).

26. W. P. Olson, G. Bethel, and C. Parker, Prep. Biochem. 7: 333 (1977).

27. R. J. Hawker and L. M. Hawker, Lab. Practice 24: 805 (1975).

28. T. Somer, Acta Med. Scand., Suppl. 456 (1966).

29. P. L. Zuideveld, "Klaring van Vloeistoffen met Behulp van Filter-poeders,: Ph.D. Thesis, Technische Hogeschool, Delft, The Netherlands (1976).

30. A. Valeri, G. Gassel, and G. Genna Experientia 15: 1535 (1979).

31. J. D. Henry and R. C. Allred, <u>Dev. Industr. Microbiol.</u> 13: 177
     (1972).
32. G. B. Tanny, D. Mirelman, and T. Pistole, <u>Appl. Environ. Micro-
     biol.</u> 40: 269 (1980).
33. M. C. Porter and W. P. Olson, U. S. Patent 4,178,248,
     December 11, 1979.

PURIFICATION OF POLYMERIC DYES BY ULTRAFILTRATION

Anthony R. Cooper, David P. Matzinger, and
Robin G. Booth

Dynapol
1454 Page Mill Road
Palo Alto, CA 94304

INTRODUCTION

Ultrafiltration is a widely accepted technique for the purification of polymeric or colloidal materials. We have previously reported[1] on the purification of a water soluble polymeric dye which contained water soluble organic impurities. The preparation of another class of polymeric dyes has been described [2] which are water soluble but the organic impurities are water-insoluble. The organic impurities are capable of bonding hydrophobically to the polymeric dye in aqueous solution, which causes special problems in their purification[3] and analysis[4]. The requirement for high purity arises from their intended use as food additives[5,6]. This work describes the development of a purification method for these polymeric dyes by ultrafiltration in mixed aqueous solvents.

EXPERIMENTAL

Materials

The preparation of this polymeric dye from a copolymer of aminoethylene and sodium ethylene sulfonate, PAE·SES and 4-bromo, 2 methyl,3'-carbethoxy anthrapyridone (Br·CMP) has been reported[2]. PAE·SES was polymerized in - house, and Br·CMP was purchased from Sandoz, Lot #ZP408. The structures are shown in Figure 1. Other impurities in these reaction mixtures were 4-amino and 4-hydroxy-2 methyl-3'-carbethoxy anthrapyridone, NH_2·CMP and OH·CMP.

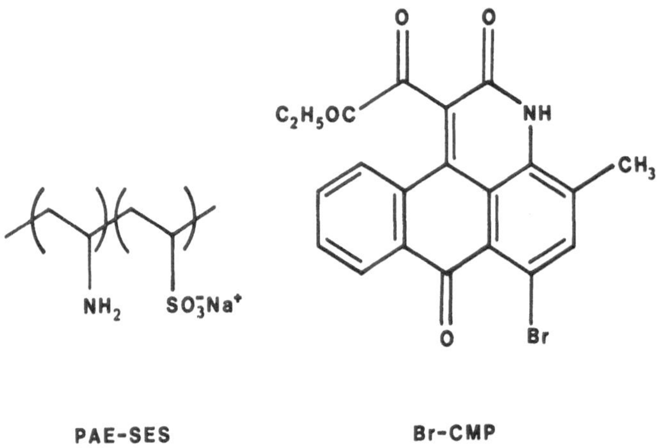

PAE-SES   Br-CMP

Figure 1.   Structure of polymeric backbone, PAE·SES and chromophore
            Br·CMP used to synthesize the polymeric dye.

OH·CMP was isolated in-house from a typical polymer reaction mixture,
and NH$_2$·CMP was synthesized in-house.

Methods

   Solubility experiments.   Tared samples in a known volume of
solvent were tumbled overnight at 25.5 ± 0.5°C.   Calibration was
performed at appropriate wavelengths and solvent compositions, since
the visible spectra vary with pH.

   Membrane stability testing.   PM10 ultrafiltration membrane
(Amicon Corp.) was tested for stability in various solvents by
soaking at least three hours.   The solvent was then washed out and
a 0.1 w/v% polymeric dye solution was used to test for flux and % dye
retention during a 50% volume reduction ultrafiltration.

   Dye stability.   The polymeric dye was tested for stability by
incubation for >50 hours at room temperature with various basic mixed
aqueous solvent systems.

   Ultrafiltration.   Steady state concentrations of components in
the initial retentate R_0 and ultrafiltrate U_0 were measured.   Diafiltration and concentration mode experiments were performed.   Flux
rates were obtained for the various solvent systems employed.   The
apparatus used employed Romicon hollow fibers and is shown in figure
2.   A recirculation pump together with control valves allowed operation at defined inlet pressures P_i and outlet pressure P_o.   An
in line heat exchanger permitted temperature control.

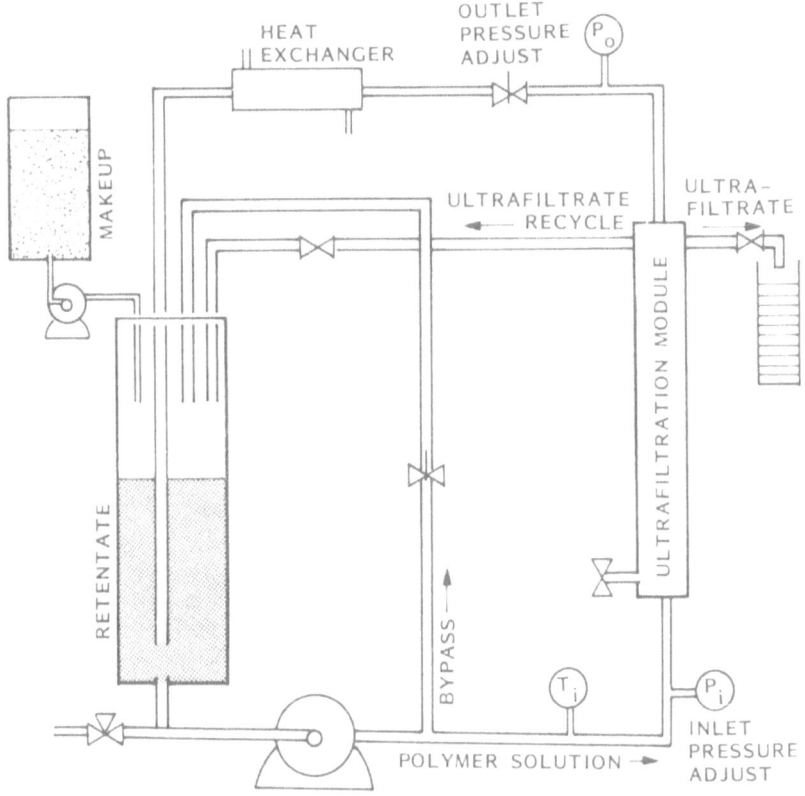

Figure 2.   Schematic diagram of ultrafiltration apparatus.

Chromatographic analysis.   Polymer and impurity concentrations
in retentate and ultrafiltrate streams were determined by gel permeation chromatography.   The conditions were:

Column: Two 0.9 cm i.d.   x 25 cm
Packing: Sephadex G-25
Solvent: 5 v/v% pyridine, 95% v/v%
                                 (0.1M phosphate, pH 12.0)
Flow rate: 1.3 ml/min
Detection: 340 nm
Injection volume: 0.2 ml
Injection concentration: CMP derivatives 5-60 ppm
                                 Polymeric dye 50-700 ppm
The calibration curves are shown in Figure 3.

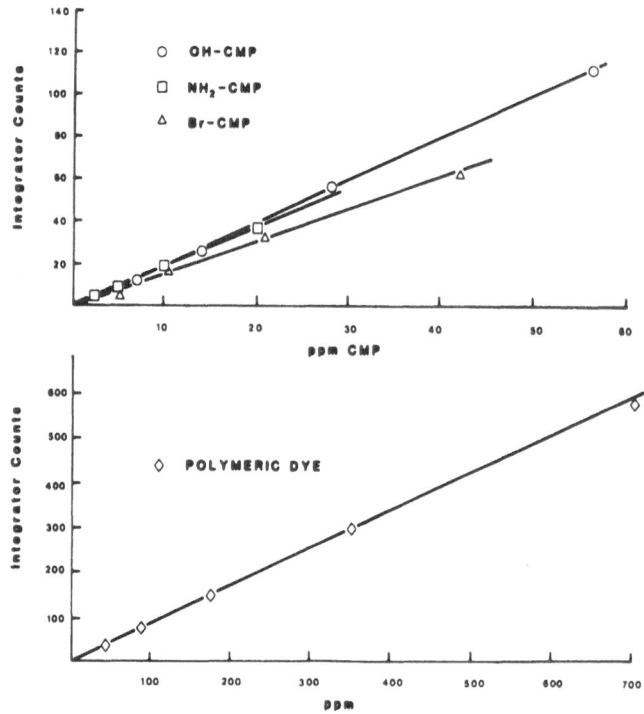

Figure 3.   Sephadex G-25 calibration curves.

RESULTS AND DISCUSSION

Solubility Experiments

    The results of the solubility determinations are shown in Table
I.   For OH·CMP, very low solubilities are obtained in neutral aqueous
solvents.   Mixed aqueous solvents are also poor solvents below pH 11.
Appreciable solubilities are obtained only at high pH (> 12.0).   At a
given pH, pyridine in the range 2-5% increases solubility.   At equivalent pH, NaOH conferred greater solubility than NH_4OH.

    For Br·CMP, the highest solubility was obtained in 5% pyridine,
pH 12.0 (NaOH).   For this compound, little solubility was achieved in
the absence of an organic cosolvent.

Membrane Stability Testing

    The times to flow 2 ml of 0.1 w/v% polymeric dye solution through
a 12.5 mm PM10 membrane disc at 35 psig after soaking the membrane in
various solvents for at least 3 hours,.are shown in Table II.   No dye
passed through the membrane, and the times were judged to be sufficiently similar that no damage to the membrane had occurred.

Table I. Solubilities of Anthrapyridone Derivatives in Mixed Aqueous
          Solvents.

SOLVENT SYSTEM	SOLUBILITY AT 25°C, ppm	
	OH·CMP	Br·CMP
50% EtOH	4	1
50% MeOH	<1	<1
50% IPA	9	1.5
5% pyridine	<1	<1
H$_2$O, pH 11.0 (NaOH)	6	<1
H$_2$O, pH 11.0 (NH$_4$OH)	17	9
50% IPA, pH 11.0 (NaOH)	87	52
50% EtOH, pH 11.0 (NaOH)	41	62
5% pyridine, pH 11.0 (NaOH)	158	43
10% pyridine, pH 11.1 (NaOH)	148	43
H$_2$O, pH 12.0 (NH$_4$OH)	102	22
H$_2$O, pH 12.5 (NH$_4$OH)	379	29
H$_2$O, pH 12.0 (NaOH)	328	5.6
H$_2$O, pH 12.5 (NaOH)	>1640	16
2% pyridine, pH 11.9 (NH$_4$OH)	254	31
2% pyridine, pH 12.0 (NaOH)	872	56
5% pyridine, pH 12.0 (NH$_4$OH)	>900	112
5% pyridine, pH 12.0 (NaOH)	>580	180

Table II. Stability Testing of Romicon PM10 Membranes.

SOLVENT SYSTEM	CONCENTRATION TIME	POLYMERIC DYE TRANSPORT
H$_2$O	42 min.	0
50% EtOH, pH 12 (NaOH)	45 min.	0
50% IPA, pH 11 (NH$_4$OH)	36 min.	0
5% pyridine, pH 12.5 (NaOH)	35 min.	0

Polymeric Dye Stability

    The results for the stability of the dye in mixed aqueous and/or
basic solvents are shown in Table III. The amount of OH·CMP or
NH$_2$·CMP were found to be essentially identical before and after
stress.

Table III.   Stability of polymeric dye in various solvents at room
             temperature.

SOLVENT	TIME HRS.	OH·CMP WT%	NH$_2$·CMP WT%
H$_2$O, pH 7	50	0.31	0.22
H$_2$O, pH 11 (NaOH)	50	0.25	0.17
50 v/v% EtOH, pH 11 (NaOH)	50	0.30	0.23
50 v/v% IPA, pH 11 (NaOH)	50	0.27	0.13
5 v/v% pyridine, pH 11 (NaOH)	50	0.27	0.16
10 v/v% pyridine, pH 11 (NaOH)	50	0.24	0.16
5 v/v% pyridine, pH 12 (NH$_4$OH)	50	0.26	0.16
0 v/v% pyridine, pH 12 (NaOH)	163	0.27	0.18
5 v/v% pyridine, pH 12 (NaOH)	163	0.29	0.20
10 v/v% pyridine, pH 12 (NaOH)	163	0.26	0.16

Ultrafiltration Results

Solvent flux rates.   Table IV gives the pertinent fluxes for
polymeric dye diafiltration experiments with mixed aqueous solvents.
The following conditions were employed:

Polymeric dye 0.9 w/v% solution
Romicon PM 10 hollow fiber module 1.3 ft^2
Inlet pressure 25 psig, outlet pressure 10 psig
Temperature 27°C

The final flux after neutralization and eight diavolumes (DV) with DI
water make-up is also given.   The results show that 5% pyridine
solvents give significantly higher fluxes than alcohol cosolvents.
The fluxes before and after the mixed solvent composition suggest
that no permanent changes in the membrane occurred.

Rejection Coefficients σ

For efficient purifications the rejection coefficient for the
polymeric dye should be close to unity, otherwise product losses will
occur.   For the impurities the rejection coefficients should approach
zero in order to minimize the number of diavolumes required.   The
rejection coefficients determined from the diafiltration experiments
are shown in Table V.   The rejection coefficients were calculated
from

$$\sigma = \frac{N + \ln\theta}{N - \ln\phi}$$

where
      $\sigma = 1 - C_U/C_R$, the rejection coefficient

Table IV. The Effect of Solvent Composition on Ultrafiltrate Flux.

RUN NO.	INITIAL SOLVENT	FLUX GSFD	DIAFILTRATION SOLVENT (6 DV)	FLUX GSFD	TERMINAL SOLVENT AFTER 8 DV WATER	TERMINAL FLUX GSFD
1	H_2O	27.6[a]	50% EtOH, pH 11 (NaOH)	15.0	H_2O, trace EtOH	19.2
2	H_2O	–	50% IPA, pH 11 (NaOH)	14.1	H_2O, trace IPA	24.8
3	H_2O	34.0	5% pyridine, pH 11 (NaOH)	53.7	H_2O, trace pyridine	35.8
4	H_2O	38.2	5% pyridine, pH 12 (NH_4OH)	55.3	H_2O, NH_4Cl, trace pyridine	21.6[b]

a Water flux prior to experiment, 103 GSFD.

b Water flux after backwash utilizing 5% pyridine, pH 11 (NH_4OH) solvent was 163 GSFD.

Table V. Rejection Coefficients for Impurities and Polymeric Dye from Diafiltration Experiments.

RUN	SOLVENT	σ OH·CMP	σ Br·CMP	σ NH$_2$·CMP	σ POLYMERIC DYE
1	50% EtOH, pH 11 (NaOH)	·∿1.0	0.56	0.79	0.987
2	50% IPA, pH 11 (NaOH)	0.88	0.47	0.75	0.951
3	5% pyridine, pH 11 (NaOH)	0.91	0.48	0.77	0.987
4	5% pyridine, pH 12 (NH$_4$OH)	0.50	0.10	0.61	0.961

N = number of diavolumes performed in constant volume mode

$\theta = C_R^N/C_R^O$, retentate concentration ratio

$\phi = V_R^N/V_R^O$, the volume reduction ratio.

For Runs 1, 2, and 3, N = 6 and ϕ = 1. For Run 4, N = 5 and ϕ = ½.
The results indicate that the polymeric dye is adequately retained by
the membrane. The only run exhibiting the desired low rejection coefficients for anthrapyridones was Run 4, the only run at high pH, viz
12.0.

Chromatograms of the unpurified polymeric dye and the final retentates from Runs 3 and 4 obtained with the Sephadex G25 column system are shown in Figure 4. The effect of higher pH in Run 4 at 5%
pyridine levels is significant. In order to investigate the molecular
weight of the polymeric dye in the ultrafiltrate, another gel Sephadex
G100 was employed. Some typical results are shown in Figure 6. The
polymeric dye in the retentate is eluted at the void volume, which was
determined using blue dextran. The ultrafiltrate contains polymeric
dye of a much lower molecular weight. Tartazine (MW 534) was used as
an elution volume marker for low molecular weight compounds. Figure
5 shows that the anthrapyridone impurities elute after this marker and
are not resolved into the components using the Sephadex G100 gel.

Further experiments were performed to investigate in more detail
the effect of pH at low pyridine concentrations, 0-5 v/v%. These experiments were performed in concentration mode; 500 ml of solution
containing 4.57 g of polymeric dye was concentrated to 250 ml. The
rejection coefficients were calculated from the equation
$$\sigma = \ell n[1/\phi + (1-1/\phi)(\bar{C}_U/C_R^O)]/\ell n(1/\phi)$$
$\bar{C}_U$ is the species concentration in the ultrafiltrate collected

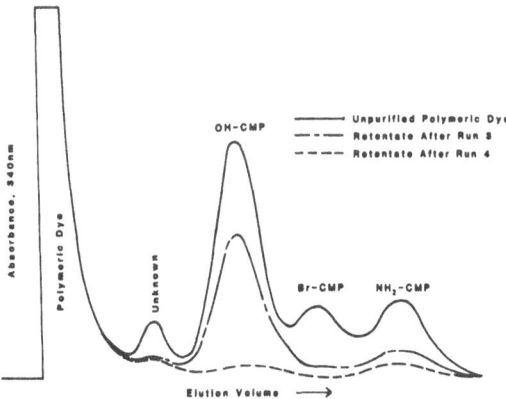

Figure 4. Chromatograms of polymeric dye retentates using Sephadex
G25 GPC columns.

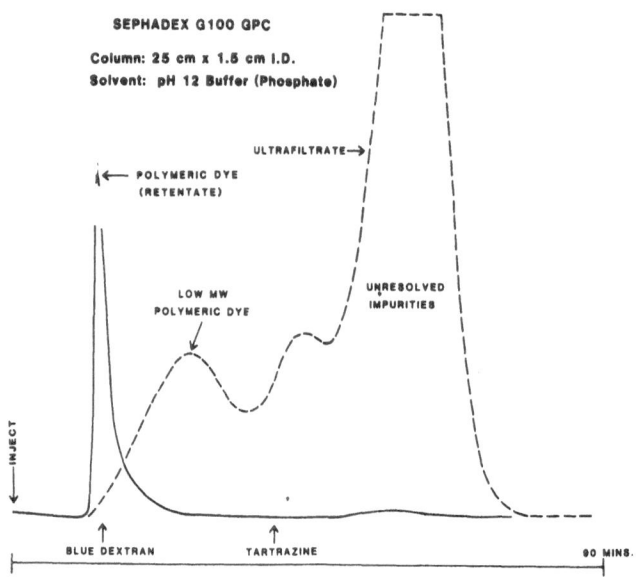

Figure 5.   Sephadex G100 GPC of retentates and ultrafiltrates.

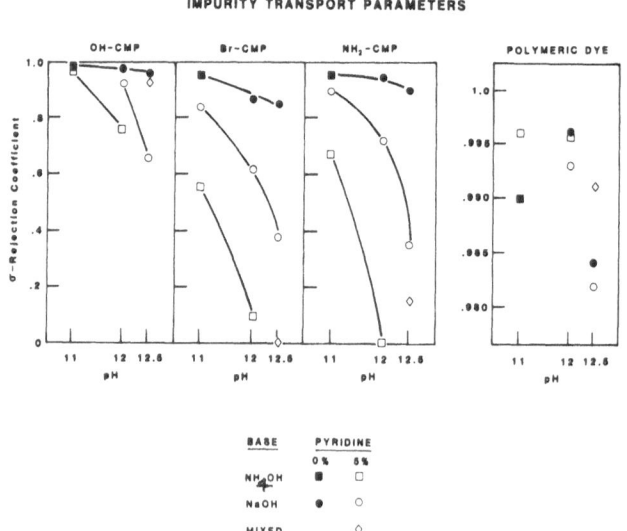

Figure 6.   Rejection coefficients as a function of pH and pyridine
             concentration.

Table VI. Calculated Values Of Rejection Coefficients For Anthrapyridones And Polymeric Dye For Various Mixed Solvents Using PM10 Hollow Fiber Module At 22°C.

RUN	SOLVENT COMPOSITION	σ OH·CMP	σ Br·CMP	σ NH$_2$·CMP	σ POLYMERIC DYE
5	0% pyridine, pH 12 (NaOH)	0.98	0.87	0.95	0.996
6	5 v/v% pyridine, pH 12 (NaOH)	0.93	0.62	0.73	0.993
7	0% pyridine, pH 12.5 (NaOH)	0.96	0.85	0.90	0.984
8	5 v/v% pyridine, pH 12.5 (NaOH)	0.66	0.38	0.35	0.982
9	0% pyridine, pH 11 (NH$_4$OH)	0.99	0.96	0.97	0.990
10	5 v/v% pyridine, pH 11 (NH$_4$OH)	0.97	0.55	0.67	0.996
11	5 v/v% pyridine, pH 12 (NH$_4$OH)	0.76	0.09	0	0.996
12	5 v/v% pyridine, pH 12.5 (mixed base[a])	0.95	0	0.18	0.991

σ calculated from $\sigma = \ell n[1/\phi + (1-1/\phi)(\bar{C}_U/C_R^O)]/\ell n(1/\phi)$.

[a] pH 12 (NH$_4$OH) + pyridine, adjusted to pH 12.5 (NaOH).

For the anthrapyridone derivatives, the transport is poor in the
absence of pyridine. At increasing pH, the transport improves drastically, NH$_4$OH appears to be more efficient than NaOH.   The Br·CMP and
NH$_2$·CMP rejection coefficients approach zero (ideal transport). However, even at the highest pH values used, the transport of OH·CMP was
still rather inefficient, but sufficiently low for the purification
to be practical. The polymeric dye showed a definite increase in
transport as the pH increased, and seemed to be insensitive to
pyridine content.

CONCLUSIONS

Ultrafiltration with mixed aqueous solvents at high pH has been
developed as a purification method for the purification of polymeric
dyes. The polymeric dye and the ultrafiltration membranes were found
to be stable under the conditions empolyed.

REFERENCES

1.   A.R. Cooper and R.G. Booth, J. Appl. Polym. Sci., **23**, 1373 (1979).

2.   D.J. Dawson, K.M. Otteson, P.C. Wang, and R.E. Wingard Jr., Macro-
     molecules, **11**, 320 (1978).

3.   A.R. Cooper, R.G. Booth, and D.P. Matzinger, US Patent 4,088,572,
     May 9, 1978.

4.   A.R. Cooper and D.S. VanDerveer, Recent Advances in Size Exclusion
     Chromatography, Advances in Chemistry Series A.C.S. Washington
     D.C. Ed., T.E. Provder, In Press.

5.   N. Weinshenker, Polymeric Drugs, L.G. Donaruma and O. Vogl,
     Eds., Academic Press (1978) p.17.

6.   W.J. Leonard Jr., Midland Macromolecular Monographs Vol. 5,
     Polymeric Delivery Systems, R.J. Kostelik, Ed., Gordon and Breach
     Science Publishers, Inc. (1978) p. 269.

PHYSICO-CHEMICAL CHARACTERIZATION OF

ENZYME-LOADED CELLULOSE ACETATE MEMBRANES

W. Pusch

Max-Planck-Institut für Biophysik
Kennedy-Allee 70
6000 Frankfurt am Main
Germany

S. Kato
Department of Physiology
School of Medicine
Yamaguchi University, U b e, 755
Japan

INTRODUCTION

Transport phenomena occurring in biological membrane systems are not only correlated with differences of the chemical and/or electrochemical potentials of the solutes and solvent across the membranes but rather with chemical reactions taking place at the membrane surfaces or within the membranes. The chemical reactions governing or affecting transport across biological membranes are essentially catalyzed by enzymes attached to the membrane matrices. For that reason, biological membranes contain immobilized enzymes and/or complete enzyme systems (organelles). The transport phenomena controlled by enzyme reactions are sometimes termed "active transport". In this connection, it should be noted that by far not all transport phenomena coupled with chemical reactions deserve the term "active transport" [1]. Most of those transport phenomena correlated with a chemical reaction rather ought to be named "facilitated or coupled transport" [2,3,4,5]. Nevertheless, it is appropriate to use synthetic membranes, containing immobilized enzymes, in order to model transport phenomena occuring in biological systems coupled with enzyme reactions. Several authors have previously reported on such systems [6,7,8,9]. Moreover, enzyme-loaded membranes are also used to prepare specific electrodes such

as enzyme and immunoresponsive electrodes, for instance, which are
employed in health control and environmental protection, for example [10,11,12].

If the effects of immobilized enzymes on the transport properties of a synthetic membrane are desired to be characterized, a
complete physico-chemical characterization of the transport and
equilibrium properties of the corresponding matrix membrane will
be wanted. Recently, the electrokinetic phenomena appearing with
asymmetric' and homogeneous cellulose acetate (CA) membranes were
reported [13,14,15]. For this reason, the preparation of urease-
containing homogeneous CA membranes suggested itself.

The electrokinetic phenomena of urease-containing CA membranes
(UCA) are not only governed by the fixed charges of the CA matrix
membrane but are also affected, in addition, by the ionogenic groups
of the enzyme attached to the membrane matrix. Therefore, the fixed
charge concentration, C_X for instance, of the UCA membranes depends
strongly on the pH and the electrolyte concentration, in general,
of the external solutions. This paper aims thus at the quantitative characterization of the pH, electrolyte concentration, and
urease catalyzed urea reaction on the membrane potential and membrane resistance of UCA membranes. The integrals of the Nernst-
Planck equations, given by Schlögl [16] for the membrane potential
and resistance, were used to perform a semi-quantitative calculation of the corresponding quantities and thus estimating the fixed
charge concentration of the membrane and the solute or ion diffusion coefficients, respectively, within the membrane.

EXPERIMENTS

Membrane Preparation

Using BAYER Cellit K-700, a CA material containing 39.1 wt-%
acetyl, casting solutions were prepared as follows. First 500 mg
of CA were dissolved in 10 ml acetone. Second, an aqueous solution
of water soluble lyophilized urease (10 wt-%; Merk, Darmstadt,
Germany) and glutaraldehyde (0.1 wt-%) was prepared and buffered
by phosphate to maintain a constant pH, pH = 7. Thereafter, 0.2 ml
of the urease solution were added under vigorous stirring to the
CA/acetone solution yielding the final casting solution for UCA
membranes after having been vigorously stirred for another 24 hrs
at 2°C. Homogeneous UCA membranes were then cast, using this
solution, on an appropriate glass plate at room temperature employ-
ing a doctor's knife of 300 µm clearance. The as-cast membranes
were first dried at room temperature for 1 hr and for another 24
hrs at 2°C. The CA reference membranes were case from the CA/
acetone solution without adding urease. The finished membranes

were throughly washed using 1 NaCl solution and finally distilled
water to remove all soluble material adhering at the membrane.
Prior to each experimental run, the finished membranes were always
conditioned storing them in the solution to be used overnight.

Measurements

Using only CA membranes of about 10 μm thickness, the partition coefficient, $K_s = C_s/c_s$, and the permeability of urea,
$P_s = K_s D_{sm}$, were measured at 25°C where C_s and c_s are the substrate
(urea) concentration within the membrane and the external solution,
respectively, and D_{sm} is the diffusion coefficient of urea within
the membrane (cm^2/s). It was then assumed that the K_s and P_s values
of UCA membranes are the same as those of pure CA membranes as it
is impossible to measure the corresponding coefficients with UCA
membranes. In addition, the permeability of the product was determined using $(NH_4)_2CO_3$ (Merk, Darmstadt, Germany) as an analog of
the product formed by the urease catalyzed urea decomposition within the UCA membrane.

The enzyme activity of urease within the UCA membrane was
characterized by measuring the rate of production of ammonium
carbonate. For that reason, an UCA membrane sample of 3.14 cm^2
effective membrane area was mounted into the partition wall of a
modified dialysis cell described in detail elsewhere [17]. The two
compartments, separated by the membrane, were then filled with distilled water or buffer solution, respectively. At time t = 0, an
equal amount of urea was simultaneously added to both compartments
and the variation of the conductivity, χ, in each compartment
followed. The conductivity increase of the external solutions is
a measure of the production rate of $(NH_4)_2CO_3$ in a quasi-steady
state of the membrane system. Figure 1 reproduces the time course
of the $(NH_4)_2CO_3$ concentration measured by both the conductivity
and the Berthelot method. The agreement between these two analytical methods is excellent. Concentrations of less than 0.7 mmol/
liter $(NH_4)_2CO_3$ can be analysed by the conductivity variation
corresponding to a conductivity change of about 50 $(\Omega cm)^{-1}$.

Furthermore, the water content, w, of the UCA membranes was
also measured. Table 1 summarizes the experimental results obtained
with UCA and CA membranes. Table 1 contains also the Michaelis-
Menton constants, K_m and V_m, characterizing the enzyme activity of
the UCA membranes. The constants were calculated from the corresponding Lineweaver-Burk plot.

RESULTS AND DISCUSSION

Using UCA membranes and NaCl solutions on both sides of the

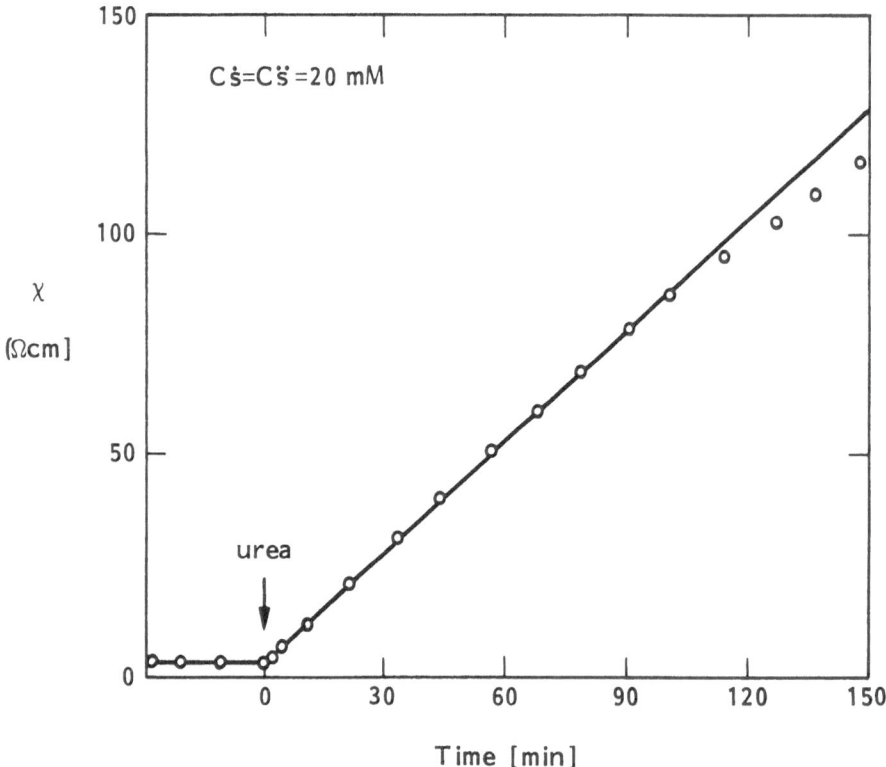

Fig. 1.   Variation of the conductivity, χ, of the external solu-
          tions with time, t, while 1 ml of a 0.5 mol/l urea solu-
          tion was added to both compartments of the dialysis cell
          at 25°C (compartment volume $\simeq$ 250 ml each).

membrane, adjusting different but constant pH values each time,
the membrane resistance, R_m, and the membrane potential, $\Delta\phi$, were
measured with and without urea present in the external salt solutions. Measuring R_m, the NaCl concentration was the same on both
sides of the membrane, $c_e' = c_e''$, whereas determining $\Delta\phi$, different
concentrations were applied keeping the ratio c_e'/c_e'' constant. As
is obvious from Figure 2, $\Delta\phi$ of the UCA membrane depends strongly
on the pH and the electrolyte concentration of the external solutions. This is a consequence of the dissociation equilibria of

Table 1.   Characteristic Parameters of CA and UCA Membranes

c_i (mol/1)	$P_s \cdot 10^8$ (cm^2/s)	$P_p \cdot 10^8$ (cm^2/s)	K_s	w (wt-%)	δ (μm)
		CA Membrane			
0.01	2.40	6.38	--	17.1	13
0.10	2.46	6.68	--	17.1	13
1.00	--	--	2.8	--	13
		UCA Membrane			
Membrane sample	$\phi_p \cdot 10^9$ (mol/cm^{2}s)	K_m (mmol)	V_m (mol/cm^{3}s)	w (wt-%)	θ
UCA-1	1.37	28.8	1.05	20.4	2.7
UCA-2	0.54	33.3	0.42	20.4	1.6

s = substrate; p = product; w referred to wet membrane weight;
ϕ_p = product flux; $\theta^2 = (V_m/K_m) \cdot (\delta^2/D_{sm})$

the ionogenic groups of both the matrix membrane and the attached enzyme such as COOH-, NH$_2$-, and His-imidazolium-groups, for instance. Depending on the pH and the electrolyte concentration of the external solutions, the UCA membrane acts like a cation or anion exchange membrane, respectively, because of the different pK values and polarities of the corresponding ionogenic groups present within the membrane. At lower pH values of the external solutions, therefore, the membrane potential of the UCA membrane changes from that of a cation (COO$^-$) to that of an anion (NH$^+$) exchange membrane. This experimental finding is in agreement with recent theoretical predictions [14]. Using the integral of the Nernst-Planck equation as given by Schlögl [16] and considering, in addition, the essential dissociation equilibria of the ionogenic groups within the membrane, the membrane potential was calculated as a function of the pH and electrolyte concentration of the external solution yielding satis-factory qualitative agreement between experiment and theoretical calculations.

The effect of the addition of urea to the external solution on R_m and $\Delta\phi$ is graphically shown in Figure 3 where the time course of R_m and $\Delta\phi$ is plotted while an amount of urea was added to the external solution at time t. Urea diffuses into the membrane and is subject to decomposition yielding NH$_4^+$, HCO$_3^-$ and CO$_3^{2-}$ ions. As a consequence of the production of these ions within the membrane, the inside pH varies resulting in a change of $\Delta\phi$ and R_m. This var-iation of $\Delta\phi$ and R_m with the pH changing is characteristic of weak ion exchange membranes as was demonstrated recently [14]. The

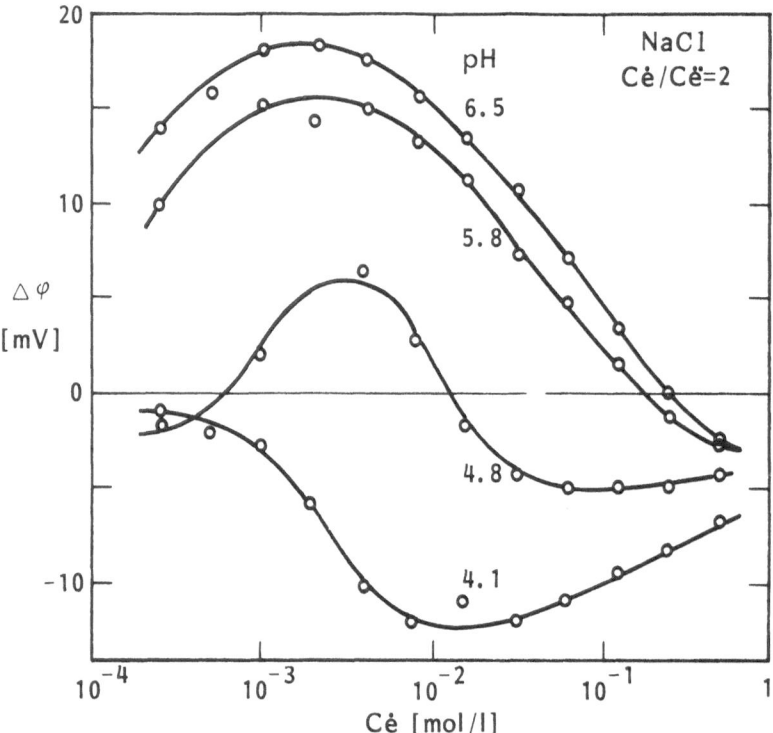

Fig. 2   Membrane potential, $\Delta\phi$, as a function of the NaCl concen-
         tration, c_e', of the external solution at 25°C and at
         different but constant pH values maintaining the ratio
         c_e'/c_e'' constant each time.

variation of R_m with increasing pH is essentially a consequence of
the variation of the pH as the resistance of the membrane is mainly
determined by the highly mobile H^+-ions. The experimental findings
manifest the strong variation of the pH within the membrane phase
originating from the production of $(NH_4)_2CO_3$.

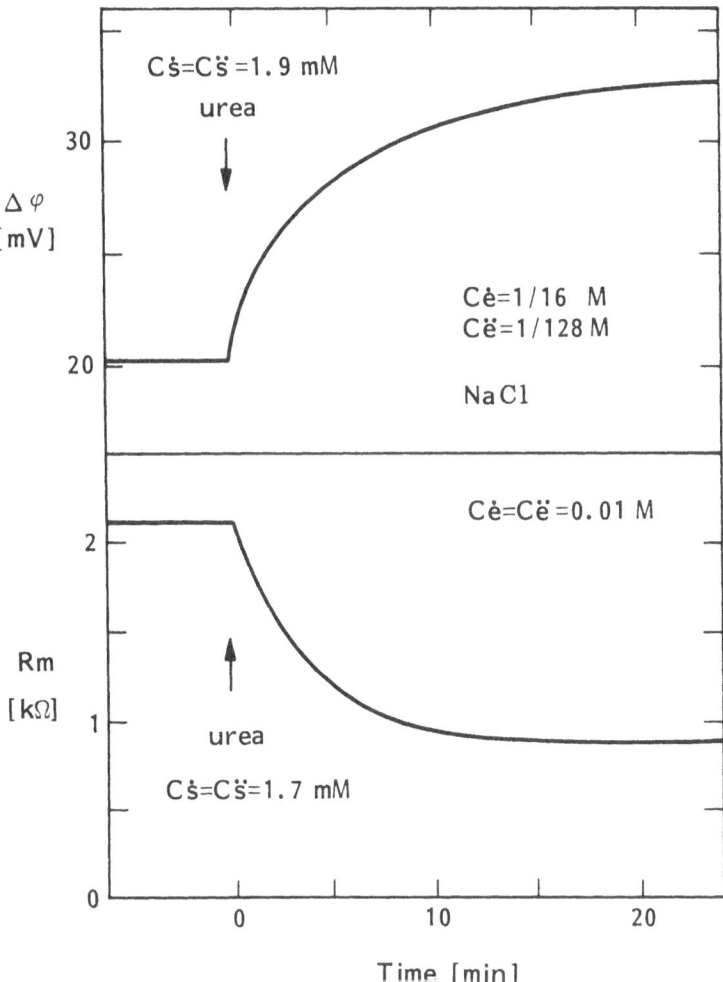

Fig. 3. Time course of the membrane potential, $\Delta\phi$, and the membrane resistance, R_m, while urea was added to the external solution at time t.

REFERENCES

1. A. Katchalsky; in: "Permeability and Function of Biological Membranes," L. Bolis, A. Katchalsky, R. D. Keynes, W. R. Loewenstein and B. A. Pethica, eds., North-Holland Publishing Co., Amsterdam, 1970, pp. 20–35.
2. W. D. Stein; "The Movement of Molecules Across Cell Membranes," Academic Press, New York, 1976.
3. W. J. Ward, III; in: "Recent Developments in Separation Science," Vol. I, N.N. Li, ed., CRC-Press, Cleveland, OH. 1972, pp. 153–161.

4. R. W. Baker, M. E. Tuttle, D. J. Kelly and H. K. Lonsdale;
    J. Membrane Sci., 2 (1977) 213.
5. E. L. Cussler; A. I. Ch. E. J., 17 (1971) 1300.
6. R. Blumenthal, S. R. Caplan and O. Kedem; Biophys. J., 7 (1967)
    735.
7. G. B. Tanny, O. Kedem and Z. Bohak; J. Membrane Sci., 4 (1979)
    363.
8. J. Meyer, F. Sauer and D. Woermann; Ber. Bunsenges. physik.
    Chem., 74 (1970) 245.
9. D. Thomas and S. R. Caplan; in: "Membrane Separation Processes,"
    P. Meares, ed., Elsevier, Amsterdam, 1976, pp. 351—397.
10. S. Kato, M. Aizawa and S. Suzuki; J. Membrane Sci., 1 (1976)
    289; 2 (1977) 39; and 2 (1977) 125.
11. M. Aizawa, A. Morioka and S. Suzuki; J. Membrane Sci., 4 (1978)
    221.
12. L. C. Clark; in: "Enzyme Engineering," L. B. Wingard, ed.,
    Wiley-Interscience, New York, 1972, p. 377.
13. H.-U. Demisch and W. Pusch J. Electochem. Soc., 123 (1976) 370.
14. H.-U. Demisch and W. Pusch, J. Colloid Interface Sci., 69
    (1979) 247.
15. H.-U. Demisch and W. Pusch; J. Colloid Interface Sci., in press.
16. R. Schlögl; Z. Phys. Chem., N.F., 1 (1954) 305.
17. W. Pusch; Desalination, 16 (1975) 65.

NOVEL WATER TREATMENT PROCESSES

WHICH INVOLVE POLYMERS

Brian A. Bolto

CSIRO Division of Chemical Technology
South Melbourne 3205
Australia

ABSTRACT

Insoluble reagents in microparticle form react at very rapid
rates. This advantage can be combined with ease of handling if the
microparticles are either bound together to form a composite
particle of conventional size held together by a polymeric matrix
which is permeable to water and the impurities present, or formed
with ferromagnetic material within them; when magnetized, the
particles flocculate strongly and quickly settle out. The magnetic
concept is particularly useful for regenerable systems which can
be operated in a continuous manner and with the use of very simple
contacting devices. Its feasibility has been demonstrated for a
number of ion-exchange reactions including one based on thermal
regeneration, for a new clarification and decolorization process
involving alkali-treated magnetite, and for a selective adsorption
process which utilizes a magnetic activated carbon for absorbing
only small molecules.

INTRODUCTION

A number of new water treatment processes have emerged from
research carried out at CSIRO, initially on the use of hot water
for the regeneration of ion exchange resins employed in desalting
brackish waters. The problems which had to be solved in the
desalting work mainly related to the need for accelerating the
rates of reaction of resins, and have led to new methods of
utilizing solid polymeric reagents in water treatment,
hydrometallurgy and the food industry.

The central theme of the work has been the exploitation of the

211

rapid reaction rates which are achieved when insoluble polymeric reagents are used in microparticle form. Because of their size, which can be as small as 1-5μm, the microparticles cause severe handling problems. When used in packed beds they provide a high resistance to the flow of water, and are prone to be lost when the bed is cleansed by backwashing. When used in conjunction with stirred vessels, they settle out very slowly in the separation stage.

To overcome the difficulty of handling such fine material, two approaches have been perfected. In the first the microparticles are bound together to form a larger composite particle, held together by a crosslinked polymeric matrix which is permeable to water and the impurities in it. This method has been used in the desalting application, when a batch mode of operation with a packed bed of resin in adequate (1).

The second technique makes use of magnetic microparticles containing small amounts of a magnetic material, such as appropriate iron oxides. When magnetized, the particles form strong flocs. On agitation, the particles disperse and react rapidly; when agitation ceases they form flocs once more and quickly settle out. Alternatively, the reagent may be demagnetized for the reaction step, and magnetized for retrieval. The magnetic concept has been applied to a variety of adsorption processes, including continuous ion exchange reactions of a number of types (2), a new continuous turbidity and colour removal process based on alkali-treated magnetite (3), and a selective adsorption process utilizing a magnetic activated carbon which can adsorb only small molecules (4). In some of these processes the reagents are present in composite particles which contain the active material, a magnetic filler and the matrix or binder.

The two approaches allow the rapid reaction rate advantage of microparticles to be combined with the handling properties of larger particles. Non-reacting forms of magnetic polymers have also been developed as recycleable filter aids (5) and as adsorbents for oil spill recovery (6).

COMPOSITE ION EXCHANGE RESINS

Thermally regenerable ion exchange has been extensively reviewed in the literature (1). A mixture of weakly basic and weakly acidic resins is employed, to give a system which is inherently slowly reacting:

$$\overline{R'CO_2H} + \overline{R''NR_2} + Na^+ + Cl^- \rightleftharpoons \overline{R'CO_2^-Na^+} + \overline{R''NR_2H^+Cl^-}$$

The formation of the charged sites at which the salt becomes adsorbed involves proton transfer from the acidic to the basic

resin. When resin particles of normal size (300 - 1200µm) are
used, the rate of salt uptake is about 100 times slower than the
rate of conventional ion exchange reactions. This has been
ascribed to the near neutral pH levels prevailing, giving an
extremely low proton concentration.

Adequately rapid reaction rates can be obtained by shortening
the path for the proton transfer step, as when both the basic and
acidic groups are present within the one resin particle. However,
should both types of groups be substituted on the one polymer
chain, low salt uptakes are experienced. The groups of opposite
character are then in too close proximity; neutralization occurs
with the formation of internal salt structures. The coupled sites
are then not available to adsorb the mobile anions and cations,
and the amount of salt adsorbed is only about 5% of that obtained
with a simple mixture of basic and acidic resins (1).

Similarly, low salt uptakes result when the groups exist on
separate polymers which are closely intertwined as in the snake-
cage resins, made by forming a flexible polymeric chain within a
crosslinked polymeric network. The rapid adsorption of salt is
negated by the low capacity of the system, now about 40% of that
of the simple mixture of resins (1).

Such internal neutralization is avoided with a mixture of the
resins, and the reaction rate can be suitably accelerated by
decreasing the size of the particles to 1-5µm. These small
particles can be readily handled by binding them together with a
water and salt permeable polymer to form a composite resin bead of
normal size which contains both the basic and acidic species (7).

A suitable polymer is poly(vinyl alcohol), crosslinked with
a dialdehyde, and present to the extent of about 40% of the
volume of the bead. The beads can be used in conventional batch
columns, adapted for thermal regeneration; a large scale plant
capable of desalting 600 Kℓ/day of a saline surface water has been
operated in Adelaide, South Australia (8).

The presence of the inert binder or matrix diminishes the
capacity of the system, makes for a complicated resin manufacturing
scheme involving three separate synthetic routes, and retards
adsorption rates slightly, since the rate determining step is now
the diffusion of ions through the matrix. Resins devoid of inert
matrix have therefore been prepared by utilizing one of the
active polymers as the binder, which envelopes the other active
polymer as shown in Fig.1.

Whilst simpler to prepare than the three component systems,
the active matrix resins do not give high performance unless steps
are taken to segregate the two types of exchange material and

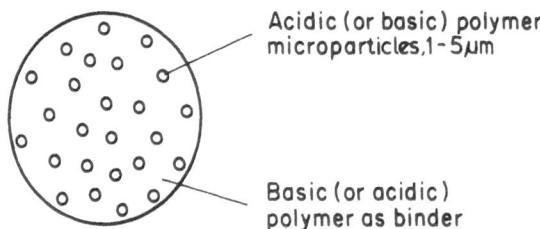

Fig.1. Structure of an active matrix resin bead, overall
        diameter 300-1200μm.

thus avoid interactions which would eliminate many sites otherwise
suitable for ion exchange.

Two Monomer Approach

The simplest, most ideal preparative route involves a one-
step synthesis from two monomers in homogeneous solution. However,
if a mixture of basic and acidic monomers such as triallylamine
and acrylic or methacrylic acid is employed, a resin is obtained
which has no thermally regenerable (TR) capacity. Salt formation
occurs prior to polymerization thereby giving a product having
the maximum possible interaction between the protonated amino and
dissociated carboxylate groups.

If the same polymerization is carried out in the presence of
added anions and cations, particularly multivalent and large
organic ions, the electrostatic interaction between the monomers
can be suppressed in favour of stronger interactions between each
monomer and a nonpolymerizable counterion (9).

After removal of the counterions, the resins have significant
TR capacities, but only about one third of the theoretical maximum,
indicating that internal salt formation has not been entirely
avoided.

If mineral acid is added to the mixture of monomers so that
the acidic monomer is uncharged, the products have no useful
capacity unless solvents other than water are employed. With the

addition of formic acid, the acidic polymer precipitates as fine
particles, around which the basic monomer polymerizes. Nevertheless,
only moderate TR capacities result, akin to those obtained via the
counterion route (9).

Better products are possible if a noncharged precursor for
the acidic groups is employed, which can be hydrolyzed after
polymerization to provide the exchange sites. Resins made from
methacrylamide and mixtures of tri and diallylamine have TR
capacities similar to those of the resins made with an inert
matrix, or about 60% of what is theoretically possible (10).

The results obtained with the two monomer approach are
summarized in Table 1. It is clear that, for polymerizations of
solutions of monomers, the chances for interaction between species
of opposite character are too high and inferior capacities are
the result.

Table 1.   Resins Prepared from Triallylamine
             and Acidic Monomers

Monomer	Blocking Method	TR Capacity 20-80°C, meq/g
Methacrylic Acid	None	0.0
Methacrylic Acid	Counterion	0.6
Methyacrylic Acid	Precipitation	0.6
Methacrylamide and Methylene-bis-acrylamide	Neutral Precursor	1.2[a]

[a] After alkaline hydrolysis; diallylamine present also.

Polymer/Monomer Route

A two-step approach is more useful, in which one of the
polymeric species is first formed as a solid particle and in a
second stage the other monomer is deliberately polymerized around
it to form the active matrix. By using the precautions outlined
in the preceding section, internal neutralization can be avoided.
Some examples are shown in Table 2, the most successful resin
being formed from divinylbenzene (DVB) crosslinked poly(ethyl
acrylate) microparticles bound together by a matrix composed of a
diallylamine/bis-diallylaminohexane (Hexa) copolymer. After
alkaline hydrolysis, the resin exhibits a TR capacity equivalent

to the target figure, indicating essentially complete utilization
of the groups present (11).

Table 2.   Resins Prepared by the Polymer/
Monomer Route

Polymeric Microparticles	Monomers to Provide Matrix	TR Capacity 20-80°C, meq/g
Polytriallylamine	Ethyl Acrylate/DVB	1.2[a]
Crosslinked Poly(Acrylic Acid)	Diallylamine/Hexa	1.2
Crosslinked Poly(Ethyl Acrylate)	Diallylamine/Hexa	2.1[a]

[a]   After alkaline hydrolysis

Other Preparative Routes

Resins of moderate TR capacity have been prepared by
photografting triallylamine onto poly(methyl acrylate) (12), by
polymerizing heterogeneous two monomer systems comprising two
immiscible solutions or a solution and a solid monomer (13),
and by substitution reactions in an existing polymeric network.
A comparison of some of the techniques has been compiled recently
(11).

MAGNETIC MICROPARTICLES

Continuous Ion Exchange

The rate advantages and accelerated settling of magnetic
microparticles were first related to ion exchange by Blesing et al.
(14) in 1970. A further property peculiar to such resins is that
in their magnetized and intensely agglomerated state they exhibit
a very high void space between the particles. A direct result of
this phenomenon is that the particles can be transported in
continuous contactors, with very little resin loss arising from
attrition. Very simple continuous contactors well suited to
large scale operation have been developed which make use of this
advantage (15,16).

Magnetic ion exchange resins have been prepared in a variety
of physical formats. The most simple is a conventional bead or
granule of resin, but of smaller than usual size, which contains
the magnetic material as a filler. One example is a weakly basic
resin made from polyamines crosslinked with epichlorohydrin (14).
Another approach makes use of the composite bead devised for
thermally regenerable resins, wherein the active resin
microparticles and magnetic filler are bound together with an
inert polymeric binder, to give a magnetic composite bead smaller
in size than normal resins. A third system is to form a small
inert magnetic polymeric core from the magnetic filler and the
binder, onto the surface of which is grafted the active polymeric
species. The graft or "whisker" resins have a further rate
advantage over normal crosslinked resins of the same small size,
in that the uncrosslinked grafts allow rapid diffusion of ions
into the active surface region, to give a 3 to 5 fold rate
enhancement. Typical dimensions are an overall bead diameter of
100-300μm, with a graft thickness of 20-40μm.

A very simple chemically initiated grafting procedure is
employed, based on hydrogen peroxide treatment of core beads of
crosslinked poly(vinyl alcohol) containing 70% by weight of a
magnetic iron oxide (17). After the peroxidation step, the
addition of ferrous ion to the monomer solution is necessary to
minimize the formation of homopolymer in solution. The amount of
grafting achieved is highly variable, depending on the nature of
the monomer, as shown in Table 3.

It can be seen that the best grafts are obtained with
acrylic acid and acrylamide, and that in general reasonable
products are possible with weakly acidic monomers. Fair results
are achieved with some basic monomers, whether strong or weak
in basic character. Strongly acidic monomers do not graft at all.

Chemical modification of the grafted chains can yield a
wider range of products (18). For example, poly(acrylic acid)
grafts can be esterified with isethionic acid, $OHCH_2CH_2SO_3H$, to
introduce sulfonic acid groups, or with epichlorohydrin and
diethylamine to introduce tertiary amino groups, again via an
ester linkage to the grafted chain. Likewise, the Hofmann
reaction with sodium hypochlorite on polyacrylamide grafts can
convert the amide groups to primary amino groups; the Mannich
reaction with formaldehyde and diethylamine introduces tertiary
amino groups connected by a methylene link to the amide nitrogen.

Magnetic ion exchange studies were first carried out on the
so-called dealkalization reaction, in which hard, alkaline
underground water is treated with a weakly acidic resin to
remove bicarbonate and an equivalent amount of hardness or other

Table 3.   Grafting to Magnetic Polymeric Cores
            of Crosslinked Poly(vinyl Alcohol)

Monomer	Amount of Graft, meq/g
Weakly acidic	
Acrylic acid	7.0
Methacrylic acid	2.5
Strongly acidic	
Sodium vinylsulfonate	0
Sodium 2-sulfoethyl acrylate	0
Sodium 2-sulfoethyl methacrylate	0
Sodium 2-sulfoethyl methacrylate	0
2-Acrylamido-2-methylpropanesulfonic acid	0
Weakly basic	
2-Dimethylaminoethyl methacrylate	0.14
2-Dimethylaminopropylacrylamide	0.15
2-Vinylpryidinium chloride	0.05
4-Vinylpyridinium chloride	1.1
2-Methyl-5-vinylpyridinium chloride	0.04
1-Ethyl-4-vinylpyridinium p-toluenesulfonate	0.5
Strongly basic	
2-Hydroxy-3-methacrylyloxypropyltrimethylammonium chloride	1.1
2-Methacrylyloxyethytrimethylammonium methyl sulfate	0.5
Non-charged precursors	
Acrylamide	6.3
Vinyl acetate	2.6
Vinylbenzyl chloride	0.2

cations:

$$2 \, RCO_2H \; + \; Ca^{2+} \; + \; 2HCO_3^{-} \rightleftharpoons \overline{(RCO_2^{-})_2 Ca^{2+}} \; + \; 2CO_2 + 2H_2O$$

Mineral acid is used to regenerate the system.

A poly(acrylic acid) whisker resin was employed which contained 50% by weight of magnetic filler, and had an overall particle size of 100–300μm. In a stirred cell, reaction with bicarbonate solution is 10 times faster than with conventional weakly acidic resins which generally have a particle size of 300–1200μm. The settling rate of the magnetized resin is three times that of conventional ion exchange resins.

With their rapid reaction rates, fast sedimentation, and ease of transport in a moving bed format, a number of continuous modes of operation become feasible for magnetic resins. Three alternative contacting systems have been investigated for the dealkalization reaction: fluidized beds, stirred tanks and pipeline reactors (15,16). The relative merits of each depend on the scale of operation.

Unfortunately, prolonged pilot studies showed that the poly(acrylic acid) whisker resin is not stable under the mildly acidic regeneration conditions. This has been ascribed to the cleavage of the grafted chains from the magnetic polymeric core, where the linking involves an ortho ester structure (19). Research is under way to overcome this difficulty. In the meantime resins containing the magnetic material embedded in small weakly acidic resin beads of otherwise conventional format are employed.

Laboratory studies have also been made of a number of other ion exchange reactions, such as the adsorption of heavy metal ions by a magnetic weakly acidic resin (20), and the removal of color from paper mill bleach plant effluents with a magnetic weakly basic resin (21).

Currently, pilot plant studies of desalination with thermally regenerable magnetic resins are receiving emphasis. It is planned to use this process to demonstrate the magnetic resin concept on a much larger scale, for the desalting of 1Mℓ/day of brackish underground water at Perth, Western Australia. A fluidized bed system is preferred in this instance.

In addition to providing a practical means of handling rapidly reacting microparticles, which means that less resin and smaller plants are required, magnetic resins allow the use of novel contacting systems. Such systems can be truly continuous in their operation, rather than intermittent as with most

commercially available units. As an added bonus, pre-clarification
is not necessary.

Colour and Turbidity Removal

Water clarification and decolorization is conventionally
carried out by forming an adsorptive precipitated floc of
aluminium or ferric hydroxide, at a pH level where the floc has a
positive charge. The impurity colloids and color molecules have
a negative charge and become bound to the floc, to be removed by
sedimentation.

A considerable effort has gone into the preparation of
regenerable magnetic adsorbents which will remove suspended solids
as well as dissolved organic material, and overcome many of the
shortcomings of the standard method, the most notable being the
slow settling properties and the formation of voluminous quantities
of sludge. Initially, research in the area was focussed on the
use of whisker resins formed by grafting a positively charged
polymer onto a magnetic polymeric core. A suitable species has
been formed by the use of 2-hydroxy-3-methacrylyloxypropyltrimethyl-
ammonium chloride (17). The whisker resin very effectively
removes colloids and color (22). It can be regenerated with brine,
but the cost of the resin relative to alum renders the process
uneconomic.

A much simpler and cheaper inorganic version of the magnetic
adsorbent has now been devised (3). It has been found that simply
treating magnetite, Fe_3O_4, of particle size 1-10µm with 0.1M
sodium hydroxide for 15 minutes gives a suitable adsorbent. The
surface charge on the activated magnetite is positive at pH 6
because of protonation:

$$\text{\textbackslash}Fe-OH + H^+ \rightleftharpoons \text{\textbackslash}Fe-OH_2^+$$

The impurities in the water adhere to the positive surface.
Collection is facilitated by magnetizing the particles, whereupon
they settle very rapidly, leaving a clear, colorless water to be
taken off from the top of the vessel. The loaded magnetite is
pumped out and regenerated with dilute alkali at pH 11.5. The
surface charge on the iron oxide is then reversed:

$$\text{\textbackslash}Fe-OH_2^+ + 2OH^- \rightleftharpoons \text{\textbackslash}Fe-O^- + 2H_2O$$

The impurities become even more negatively charged at the raised
pH, and are repelled from the magnetite. After washing, the
magnetite is demagnetized before the next clarification step.

Waters of high color and low turbidity can be successfully
treated by the activated magnetite alone, but pH levels of 4 to

5.5 may be necessary. For more turbid waters, the surface area
available is inadequate to cope with the colloids present, and
another coagulant must be added. A cationic polyelectrolyte is
the preferred reagent, as it assists in bridging the residual
turbidity to form flocs which then bind to the magnetite surface.
Weakly basic polymers are the logical choice, and provided that
they are sufficiently basic and that the charge density is high
enough, clarification can be effected at pH 6, as depicted in
Fig.2(a). Under the alkaline regeneration conditions the weakly
basic groups lose their charge, and regeneration is facilitated
as shown in Fig.2 (b).

     The optimum polyelectrolyte structure depends on the nature
of the impurities in the water. All available commercial
polyelectrolytes and many specially synthesized materials have
been studied (over 70 polymers in all). Species which are highly
charged under the clarification conditions, whether weakly basic
like polyethyleneimine or strongly basic like poly(4-vinyl-N-
methylpyridinium chloride), perform well. The weakly basic
example also results in very efficient regeneration. However,
because of the short length of these polymer chains, the impurity
laden magnetite particles are sensitive to the shear stress which
develops during the magnetization step prior to separation of the
product water. The problem can be alleviated by the use of very
long chain copolymers of acrylamide and amino acrylates.
Unfortunately, such polymers do not separate cleanly from the
magnetite under the alkaline regeneration conditions, apparently
because of a specific binding to the iron oxide. The situation
can be restored, but at a cost, by a higher dose of alkali at
about weekly intervals.

     In the short term, satisfactory clarification performance can
be achieved with a mixture of polymers: a highly charged species
of moderate chain length for good color removal and an acrylamide/
amino acrylate copolymer of very long chain length for retention
of the turbidity by the magnetite in the separation stage (23).
Regeneration difficulties are then minimized. The ultimate answer
is a polyamine of very high molecular weight; research on this
topic is well advanced.

     The process has passed through a successful pilot plant stage,
treating a highly colored and turbid underground water in Perth.
A much larger demonstration and production facility has now been
constructed which can treat 35 Ml/day.

Selective Magnetic Adsorbents

     During research on the nature of the crosslinked poly(vinyl
alcohol) used as a binder in the thermally regenerable composite
resins, and in the magnetic core of general purpose whisker resins,

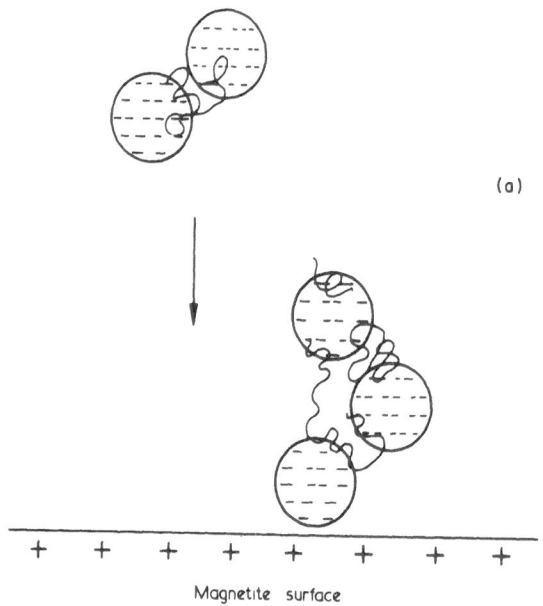

Fig.2(a)   Clarification at pH 6, when the polyelectrolyte is
           positively charged and bridges the turbidity particles.

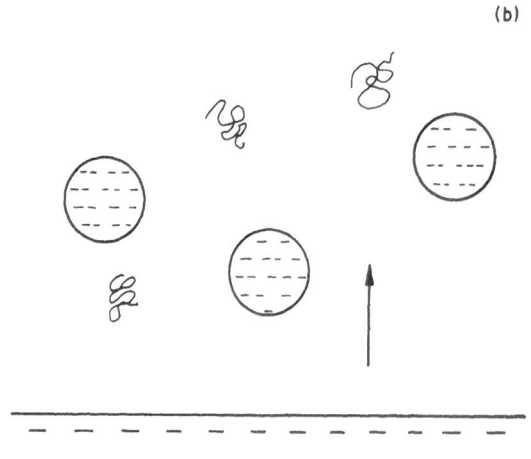

Fig.2(b)   Regeneration at pH 11.5, which removes the positive
           charge from a weakly basic polyelectrolyte and
           reverses the charge on the magnetite.

it was observed that the polymer would exclude proteins of
molecular weight larger than 12,000.

By embedding an adsorbent such as activated carbon, together
with iron oxide, within the binder, composite magnetic adsorbent
beads can be produced which will remove small nuisance molecules
from solution, but not larger species such as vitamins, proteins
and nucleic acids (4). The system is being tested in a food
industry application for the removal of bitter principles from a
yeast extract, leaving untouched the valuable macromolecules
which have a high nutritional value. The magnetic properties of
the adsorbent particles are very useful in the separation of the
loaded adsorbent from cell fragments and other insoluble material.
Regeneration is achieved with alkali or a solvent.

Further Magnetic Polymer Systems

The high void space in the magnetized agglomerates of
magnetic polymers can be put to use in two further applications.
Insoluble hydrophilic magnetic materials based on copolymers of
vinyl acetate and vinyl chloride, and made as irregular shapes
rather than spheres, function as regenerable filter aids (5).

Hydrophobic magnetic polymers can be utilized as recoverable
reagents for absorbing oil spills on harbor waters. Made as
vesiculated polystyrenes, so that they float, the magnetised
particles trap the oil within the void space, and can be picked
up by magnetic means. A craft has been designed and built which
demonstrates the process.

Other applications of the magnetic concept abound in the
biotechnology area, where the prime advantage is the easy
separation of the magnetic reagent from the mass of solid material
present, without the need for a clarification or filtration step
before separation of the desired soluble product is attempted.
Enzymes fixed on magnetic supports have been reported, as have
antibodies bound to magnetic spheres. Magnetically supported
microbes are also possible. The application of magnetic polymers
in ion exchange and other processes has been reviewed recently (24).

CONCLUSIONS

The enhanced reaction rates of microparticles can be exploited
if handling is facilitated by forming them into larger composite
beads or by implanting magnetic properties. The magnetic approach
in particular has led to new continuous contacting devices which
are applicable to a wide range of treatment processes. Less
reagent and smaller plants are required. The other important
advantages are simplicity of design and the avoidance of
preclarification when the aim is isolation of a soluble product or
contaminant.

REFERENCES

1. B.A. Bolto and D.E. Weiss, in "Ion Exchange and Solvent
    Extraction, J.A. Marinsky and Y. Marcus, Eds., Vol.7,
    Dekker, New York, p.221 (1977).
2. B.A. Bolto, D.R. Dixon, R.J. Eldridge, L.O. Kolarik, A.J.
    Priestley, W.G.C. Raper, J.E. Rowney, E.A. Swinton, and
    D.E. Weiss, "The Theory and Practice of Ion Exchange,"
    Society of Chemical Industry, London, paper 27 (1976).
3. L.O. Kolarik, A.J. Priestley, and D.E. Weiss, Proc. 7th
    Federal Convention, Australian Water and Wastewater
    Association, p.143 (1977).
4. D.R. Dixon and L. Lydiate, J. Macromol. Sci.-Chem., A14: 153
    (1980).
5. B.A. Bolto, K.W.V. Cross, R.J. Eldridge, E.A. Swinton, and
    D.E. Weiss, Chem. Eng. Progr., 71: 47 (1975).
6. H.A.J. Battaerd, B.A. Bolto, D.R. Dixon, R.J. Eldridge, E.A.
    Swinton, D.E. Weiss, and P.H. Young, J. Polym. Sci. Polym.
    Symp. Ed. 49: 211 (1975).
7. B.A. Bolto, K.H. Eppinger, A.S. Macpherson, R. Siudak, D.E.
    Weiss, and D. Willis, Desalination, 13: 269 (1973).
8. P.J. Cable, R.W. Murtagh and N.H. Pilkington, The Chem. Eng.
    624 (1977).
9. B.A. Bolto and R. Siudak, J. Polym. Sci. Polym. Symp. Ed. 55:
    87 (1976).
10. B.A. Bolto, M.B. Jackson, R.V. Siudak, H.A.J. Battaerd, and
    P.G.S. Shah, J. Polym. Sci. Polym. Symp. Ed., 55: 87 (1976).
11. B.A. Bolto, K.H. Eppinger, M.B. Jackson and R.V. Siudak,
    Desalination, submitted for publication.
12. M.B. Jackson and W.H.F. Saase, J. Macromol. Sci.-Chem., A11:
    1137 (1977).
13. M.B. Jackson, J. Macromol. Sci.-Chem., A12: 853 (1978).
14. N.V. Blesing, B.A. Bolto, D.L. Ford, R. McNeill, A.S.
    Macpherson, J.D. Melbourne, F. Mort, R. Siudak, E.A.
    Swinton, D.E. Weiss, and D. Willis, "Ion Exchange in the
    Process Industries", Society of Chemical Industry, London,
    p.371 (1970).
15. N.J. Anderson, D.R. Dixon and E.A. Swinton, J. Chem. Tech.
    Biotechnol., 29: 332 (1979).
16. B.A. Bolto, D.R. Dixon, A.J. Priestley, and E.A. Swinton,
    Progr. Water Technol., 9: 833 (1977).
17. B.A. Bolto, D.R. Dixon, and R.J. Eldridge, J. Appl. Polym.
    Sci., 22: 1977 (1978).
18. R.J. Eldridge, in "Modification of Polymers", C.E. Carraher
    and M. Tsuda, Eds., ACS Symp. Series No.121, Amer. Chem.
    Soc., Washington, p.139 (1980).
19. J.S. Bates and R.A. Shanks, J. Macromol. Sci.-Chem., A14: 137
    (1980).
20. D.R. Dixon and D.B. Hawthorne, J. Appl. Chem. Biotechnol.,
    28: 10 (1978).

21. B.A. Bolto, A.J. Priestley, and R. Siudak, _Appita_, 32: 373
     (1979).
22. N.J. Anderson, B.A. Bolto, R.J. Eldridge, E.A. Swinton and
     L.O. Kolarik, _Water Res._, accepted for publication (1980).
23. N.J. Anderson, N.V. Blesing, B.A. Bolto, D.R. Dixon, A.J.
     Priestley, and W.G.C. Raper, _Proc. 9th Federal Convention,
     Australian Water and Wastewater Association_, (1981).
24. B.A. Bolto, in "_Ion Exchange Processes in Pollution Control_,"
     Vol.2, C. Calmon and H. Gold, Eds., CRC Press, Boca Raton,
     Fla., p.213 (1979).

SYNTHESIS AND THERMAL REGENERATION OF POLYMERIC CROWN ETHERS

Abraham Warshawsky and Nava Kahana

Department of Organic Chemistry
The Weizmann Institute of Science
Rehovot, Israel

Polymeric materials carrying pendant crown-ether groups were prepared by: (i) direct polymerization of vinylbenzo crown ethers (1); (ii) condensation polymerization of dibenzo crown ethers with formaldehyde (2).

Recently, we have presented a different approach, based on a one-step insitu cyclization reaction (3). Thus, a nucleophilic substitution reaction between two electrophilic centres (benzyl-halide groups, part of the polymeric matrix) and a cation-templated polyglycol takes place, leading to large macrocycles. The so-called "polymeric pseudocrown ethers" with 7-14 oxygen ligands per cycle are able to complex transition metal anions in their conjugate-acid form.

In the present paper, we wish to extend this approach to the synthesis of polymeric crown ethers. This is accomplished by reversing the role of the functional groups and putting the nucleo-philic centres in the polymeric matrix. Thus, the alkylation of catechol with chloromethylstyrene-divinylbenzene copolymer (see Scheme 1), produces the polymer-bound catechol which upon reaction under Williamson conditions, with a series of polyglycol dihalides, affords macrocyclic polymeric ethers in fair yields.

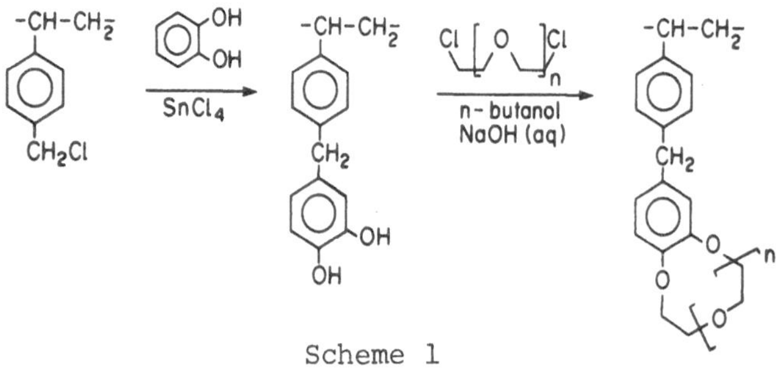

Scheme 1

The sequence of reactions described in Scheme 1 has been carried out in two parallel series. In the first, the polymers carried one substituted ring against one non-substituted aromatic ring, in the second series, one substituted aromatic ring in seven non-substituted rings. This was aimed to detect the influence of matrix variation on the ion complexation patterns.

The degree of conversion of the catechol groups into macro-cyclic ether groups and the residual concentration of diol groups were estimated from elemental analysis, weight-gain data and the analysis of dinitrophenyl derivatives of the residual diol groups. The description and properties of the polymeric crown ethers are presented in Table 1.

The synthetic transformations described in Scheme 1 were followed optically. Scanning electron microscopy photographs have revealed that the porous agglomerate structure typical to macro-porous polymers is retained. The internal structure of polymer crown-6 is shown in Fig. 1 (x 10300 magnification) and the surface in Fig. 2.

Next, the ion coordination patterns for the polymeric crowns were determined, using distribution and column techniques. Equilib-rium distribution values in the temperature range of 20-60°C, for the perchlorate, thiocyanate or bromide salts, have led to the following conclusions:

1)   The spheric recognition patterns, typical to crown ethers in solutions, are fully incorporated in the polymeric analogues as shown in Fig. 3.

2)   The highest binding constants recorded are for potassium ions, as shown in Fig. 4.

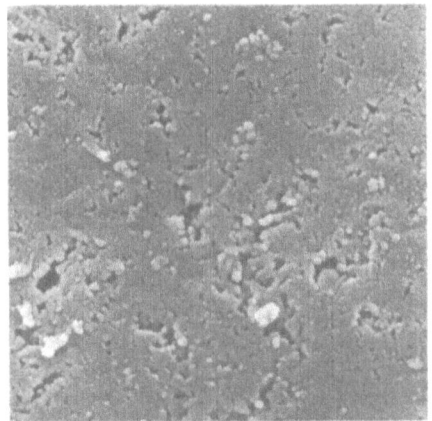

Fig. 1

Fig. 2

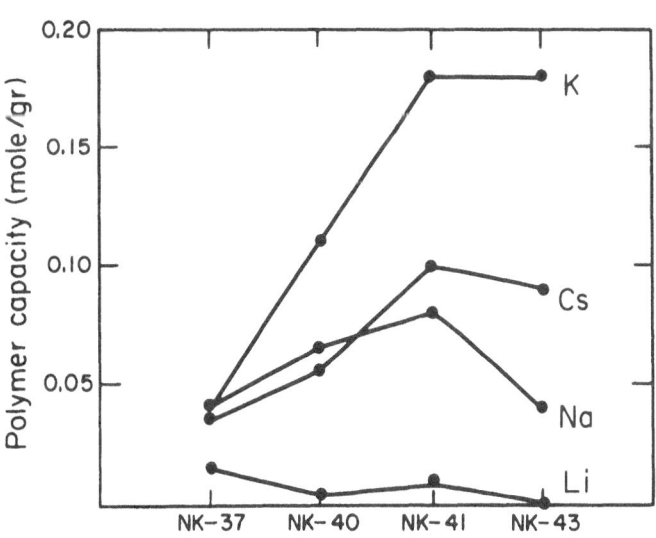

Fig. 3   Recognition patterns of polymeric crown ethers

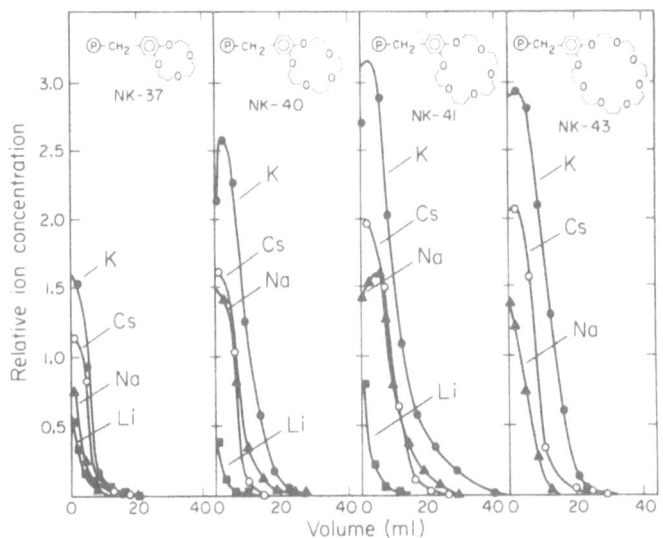

Fig. 4   Variation in ion concentration during elution at 60°C

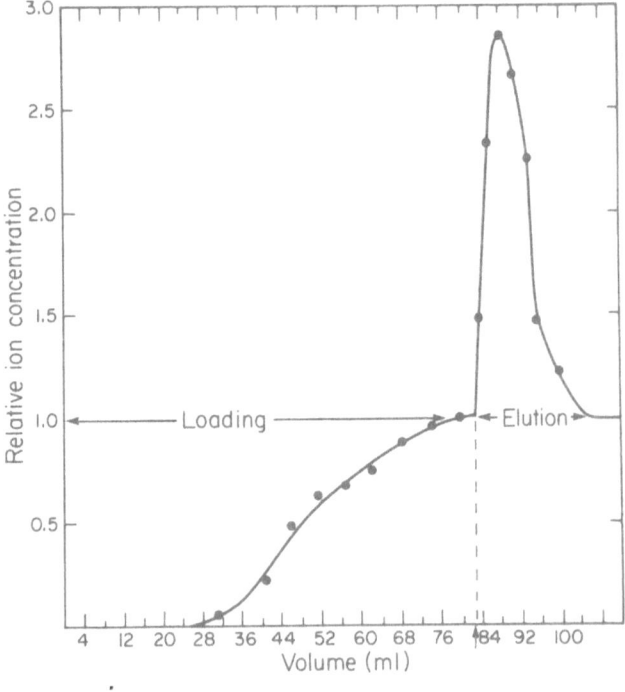

Fig. 5   Spontaneous elution by thermal shock at 60°C

Table 1.   Synthesis of Polymeric Crown Ethers by Condensation of
           Polymer-bound Catechol and Polyglycol Dihalides

Polymer	No.	Glycol dihalide	Residual chlorine (%)	Active group concentration on polymer (mmole/g)	
				Diol	Crown
Crown-4	NK-37	triethylene	1.11	0.64	1.42
Crown-5	NK-40	tetraethylene	0.48	0.61	1.28
Crown-6	NK-41	pentaethylene	0.24	0.61	1.10
Crown-8	NK-43	heptaethylene	0	0.78	0.93

Notes: 1) Weight increases of 20-27% were recorded.
           2) Cyclization yields ranged between 41 and 54%.

   3)   The polymers bind alkali metal cations by two mechanisms,
the free catechol groups by an exchange mechanism and the crown
groups by a coordination mechanism.

   4)   The coordination-type binding is inversely dependent on
temperature.

   The last phenomenon has considerable significance, as it
allows thermal regeneration of the polymer-bound cations.   In
cyclic experiments, binding was accomplished at 20°C and quantitative elution at 60°C.   In a binding experiment, sudden thermal
shock caused spontaneous elution and threefold increase in original
ion concentration (see Fig. 5).   The explanation for the thermal
regeneration phenomenon of polymeric crown ethers lies more in
enthalpy and enthropy changes caused by polymeric matrix variations
than in ordinary shifts in water dissociation constants, typical
to ion exchange resins (4).

REFERENCES

(1)   J. Smid et al., Pure Appl. Chem., 51:111 (1979).
(2)   E. Blasiuss et al., Z. Anal. Chem., 284:337 (1977), and
      references therein.
(3)   A. Warshawsky et al., J. Am. Chem. Soc., 101:4249 (1979).
(4)   B. A. Bolto and D. E. Weiss, "Ion Exchange and Solvent
      Extraction," Vol. 7, p. 221, J. A. Marinsky and Y. Marcus,
      ed., Marcel Dekker, Inc., New York (1977).

ION-SELECTIVE POLYMERIC-MEMBRANE ELECTRODES WITH IMMOBILISED

ION-EXCHANGE SITES

G.C. Corfield, L. Ebdon and A.T. Ellis

Department of Chemistry
Sheffield City Polytechnic
Sheffield S1 1WB, England

INTRODUCTION

Ion-selective electrodes (ISE) offer considerable advantages
in analytical chemistry. They are simple and inexpensive to operate, sensitive and suited to on-line measurements. Thus, in addition to their application to laboratory analyses, applications to
pollution monitoring, *in vivo* biological measurements and process
control have been proposed. It is possible to identify several
criteria which ISE should meet; such as, 'Nernstian' response over
a wide activity range, specificity for the primary ion, robustness,
long life-time and ability to function in hostile environments.

Until comparatively recent years the range of ISE was limited
to classical glass electrodes and electrodes using certain crystalline membranes. The introduction of electrodes based on liquid
ion-exchangers, e.g. that reported by Ross in 1967, greatly extended the range of available electrodes. The Ross electrode[1] made
use of calcium bis-didecylphosphate as sensor in conjunction with
dioctyl phenylphosphonate (DOPP) as solvent-mediator, both supported on a Millipore filter. The electrode was of a rather complex
design and both the limit of detection and the lifetime were limited by the leakage of the liquid ion-exchanger from the filter.
In 1970, Moody *et al*[2] reported a modification of this system in
which the liquid ion-exchanger was entangled in a PVC matrix.
Although this gave a cheap and simple ISE, the exchanger could
still be leached from the polymer matrix, again limiting the
operational lifetime of these electrodes.

In this study ion-exchange sites have been covalently bound
to a host polymer, thereby preventing the leaching process and

producing electrodes with longer life-times. The effectiveness of
this approach has been demonstrated with a calcium ISE.[3-5] Membranes were prepared by cross-linking a styrene-butadiene-styrene
(SBS) triblock copolymer with triallyl phosphate (TAP), then hydro-
lysing and conditioning the resulting membrane to give pendant
calcium dialkyl phosphate exchange groups. A comparison of this
type of electrode with commercially available types has been made,
and an application to the determination of calcium in a coking
plant effluent will be described. The effects upon analytical
characteristics of modifying the ion-exchange group and incorporating dioctyl phenyl phosphonate into the matrix have also been
investigated.

EXPERIMENTAL

Cross-linked polymers were prepared by casting a solution of
SBS, the phosphorus-containing monomer and ABIN in tetrahydrofuran,
and initiating with ultraviolet irradiation.[3] The cast elastomers
were hydrolysed in methanolic potassium hydroxide solution, when-
ever necessary, to generate the ion-exchange group. Discs, cut
from these membranes, were mounted into electrode bodies and condition in 10^{-2} M Ca^{2+} solution.

Calibrations were established using a digital millivoltmeter and
solutions of pure calcium salts in distilled de-ionised water at
25 ± 0.5 °C. Calcium ion activity was calculated from $a = c\gamma$, and
the activity coefficient from an extension of the Debye-Huckel
equation shown below:

$$-\log \gamma_{Ca^{2+}} = z^2 \left(\frac{a\sqrt{I}}{1+\sqrt{I}} - 0.2 \ I\right)$$

Potentiometric selectivity coefficients were evaluated by a
mixed solution method, with an interferent ion concentration of
10^{-3} M, and calculated using the following relationship:

$$k_{CaM}^{pot} = a_{Ca^{2+}}/(a_{M^{z+}})^{2/z}$$

RESULTS AND DISCUSSION

A poly(styrene-b-butadiene) triblock elastomer (SBS) was
chosen as the polymer because it contained the necessary unsatu-
ration for cross-linking, had mechanical behaviour similar to
natural rubber vulcanisates without requiring cross-linking,
could be dissolved in solvents and was easy to process. In order
to incorporate the phosphate exchanger, radical cross-linking
using triallyl phosphate (TAP) was carried out (Fig. 1). Membranes were prepared by ultraviolet irradiation of a solution of
SBS, ABIN and TAP in tetrahydrofuran, while allowing solvent to

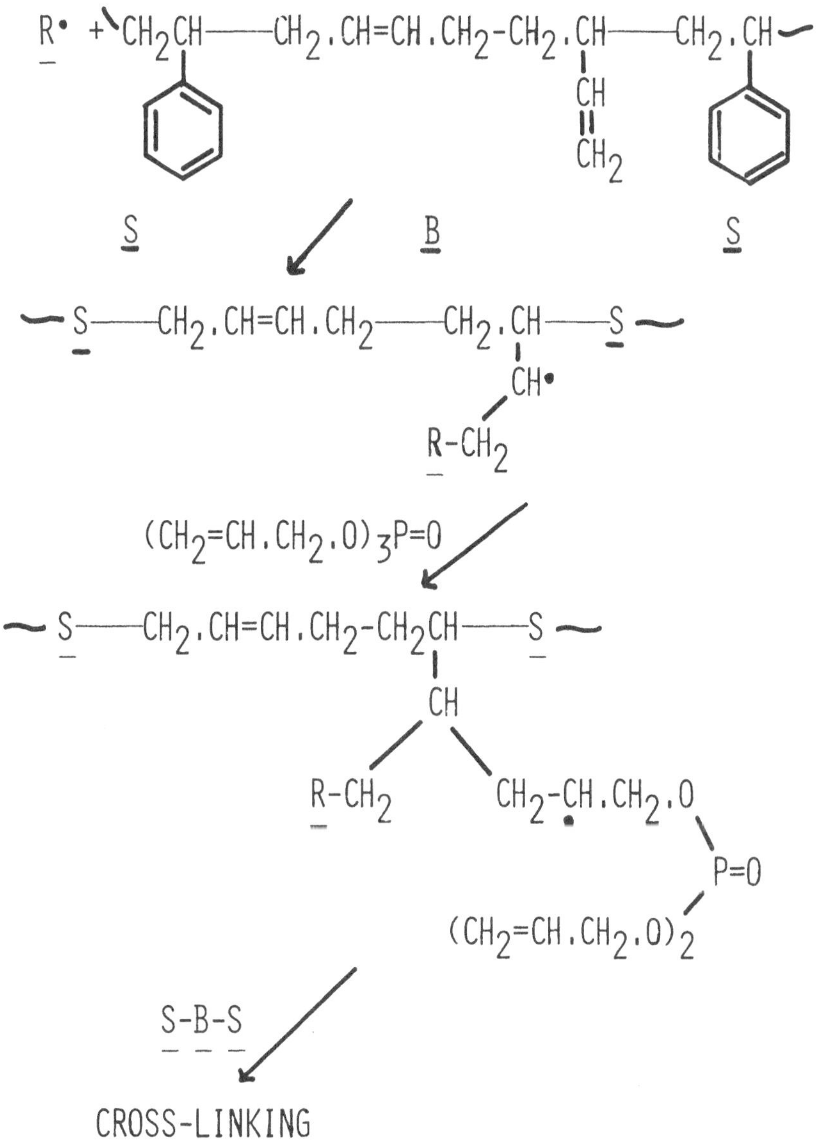

Fig. 1. Immobilisation of phosphate groups

evaporate, to form strong clear membranes about 1 mm in thickness (Fig. 2). Membranes cast containing 4.5% TAP and 2.4% ABIN were found to give the optimum performance. The incorporation of the phosphate groups by covalent bonding was shown by the determination of tetrahydrofuran-insoluble phosphorus in the membranes. To obtain the dialkyl phosphate sensor, it was necessary to hydrolyse the trialkyl phosphate system with alkali (Fig. 3). It was found necessary to balance the degree of hydrolysis and extent of cross-linking to obtain functional electrodes. Hydrolysis with 5% methanolic potassium hydroxide under reflux for 5 h. was the most suitable for membranes produced using 4.5% TAP. Discs, 10 mm in diameter, were cut from these membranes and bonded to the end of clear PVC tubing which was mounted on a modified pH electrode (Fig. 4). The mounted electrodes were soaked overnight in 10^{-2} M Ca^{2+} solution in order to replace the K^+ form of the exchanger with the Ca^{2+} form.

Electrodes formed using TAP were found to give a Nernstian calibration slope from 10^{-1} to 10^{-5} M Ca^{2+} or lower. In general, the selectivity of these electrodes was in the order $Ca^{2+} > Ba^{2+} > Mg^{2+} > M^+$, and these data are presented graphically in Fig. 5. Numerically, the k_{CaM}^{pot} values are high, but it can be seen that this does not necessarily preclude the electrode from use in a medium containing 10^{-2} M Ca^{2+} and 10^{-3} M Na^+. The effect of varying the background sodium activity is seen in Fig. 6, which shows that the electrode fails to respond to calcium ion activity at a background sodium concentration of 10^{-1} M.

Comparing this TAP electrode to the Orion 92-20[1] and a PVC calcium electrode[6] with respect to the criteria outlined earlier, the relative merits of each type can be seen (Table 1).

Table 1.   Comparison of three calcium electrodes

Characteristic	Orion	PVC	TAP
Range/M	10^{-5}-10^{-1}	10^{-5}-10^{-1}	10^{-6}-10^{-1}
Slope/mV.Decade^{-1}	30	30	30
pH range	5.5-10	5-9	4-10
Response time/s	5	2	2
Operational lifetime/weeks	4	6	24
Selectivity coefficients, k_{CaM}^{pot}			
Mg^{2+} (10^{-3} M)	0.014	0.086	0.3
Na^+ (10^{-2} M)	0.025	1.1	8
NH_4^+ (10^{-3} M)	36	0.60	76

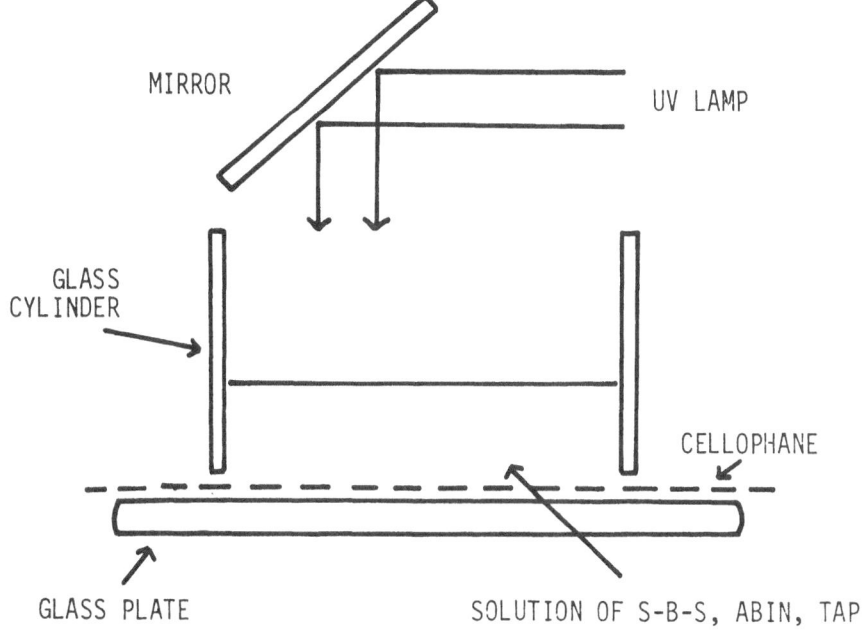

Fig. 2   Casting of Membranes

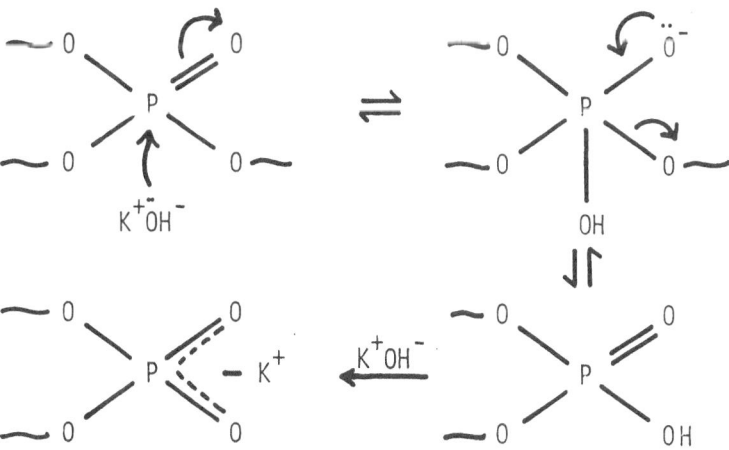

Fig. 3   Hydrolysis of trialkylphosphate groups

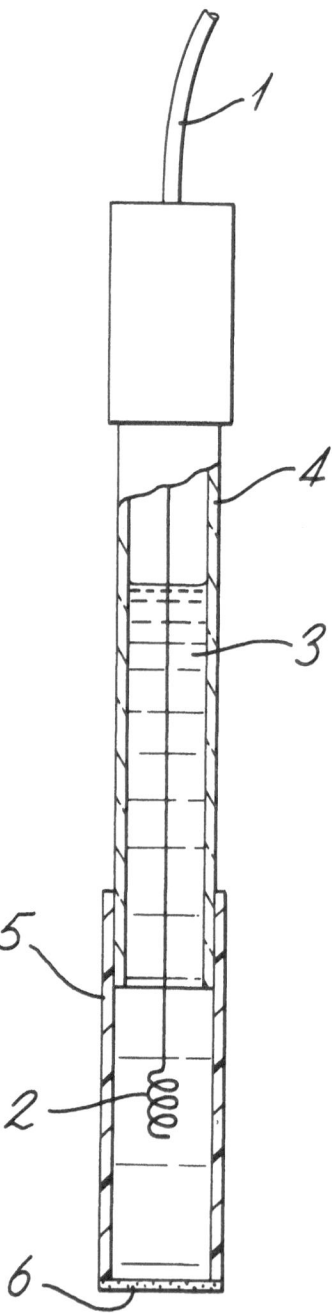

Fig. 4.   Ion-selective electrode (1) screened cable; (2) silver/
          silver chloride electrode; (3) 0.1 M aqueous calcium
          chloride solution saturated with silver chloride; (4)
          pH electrode body; (5) PVC tubing; (6) ion-selective
          polymeric membrane.

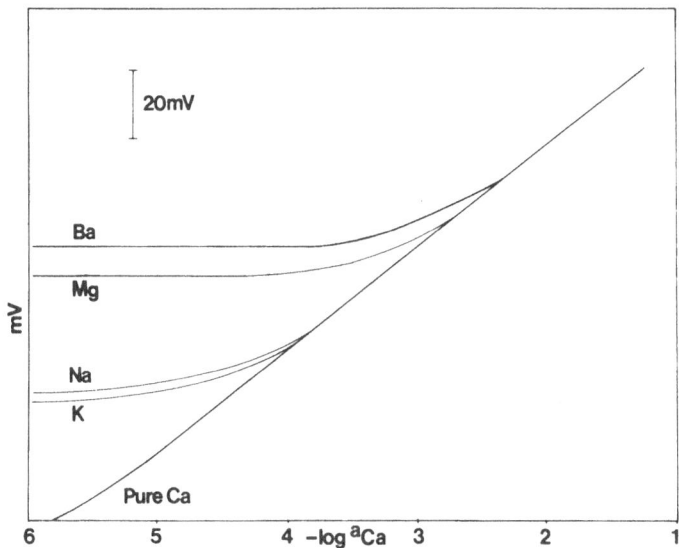

Fig. 5. Calibration graph for TAP electrode; selectivity coefficients (k_{CaM}^{pot}): Ba, 0.8; Mg, 0.3; Na, 30; K, 27 at interferent level 10^{-3} M.

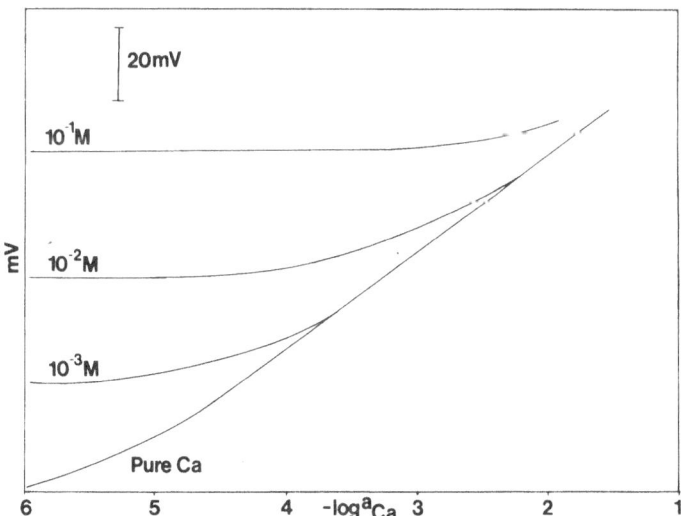

Fig. 6. Effect of sodium concentration on electrode response;
k_{CaNa}^{pot}: 30 (at 10^{-3} M), 8 (at 10^{-2}M), 0.1 (at 10^{-1} M).

In particular it is clear that the TAP electrode shows an extended
linear range and functions over a wider pH range than the other
electrodes. It exhibits a longer working lifetime, in excess of
six months, but does have poorer selectivity, particularly in the
presence of magnesium ion. This lack of selectivity may preclude
use in medical applications, but the durability of the TAP electrode should be useful in applications where selectivity is not of
paramount importance and we have demonstrated this in the monitoring of coke oven effluents.

All three electrodes have been used for the determination of
calcium in a coking plant effluent and the results compared with
values obtained by standard methods (Table 2). In the river above
the plant there is close correlation between all methods. In the
plant effluent, however, none of the electrodes, using any technique, gave results in agreement with the independent methods.
The results were high and with the Orion and PVC electrodes response
was erratic in the effluent. Rapid deterioration and discolouring

Table 2.   Coking Plant Effluent Analysis

Calcium concentration/mg.dm^{-3}

Technique/Electrode	River above	Effluent	River below
EDTA titration	108	2800	2075
AAS	114	2820	2120
ICP/AES	109	2800	2140
Direct Potentiometry			
Orion 92-20	112	5590	3850
PVC	97	3320	2410
TAP	112	3220	2525
Gran's Plot			
Orion 92-20	120	3310	2125
PVC	124	6880	2405
TAP	105	3410	2525
Standard Addition			
Orion 92-20	125	3365	2200
PVC	140	5410	3775
TAP	110	3770	2610

of the membranes of both these electrodes occured such that contact
with the effluent had to be minimised. The PVC type had to be
fitted with a new membrane after 5 weeks and the Orion was completely refurbished with ion-exchanger and a new membrane after
approximately the same period of operation. From a practical
point of view the TAP electrode is superior in this application
because it is robust, stable and has a long lifetime. The high
results for calcium obtained using the electrodes compared to the
standard methods, are considered to be due to the interfering
effect of NH_4^+ (see Table 1) in the effluent.

Despite the overall advantages of the TAP electrode, the
selectivity is inferior to that shown by other calcium ISE. In
an attempt to improve the selectivity we have modified the ion-
exchange site and investigated the effect of this upon electrode
performance. Membranes have been prepared in which the TAP used
in our original system has been replaced by tri-10-undecenyl
phosphate (TUP) or diallyl phenylphosphonate (DAPP). After hydrolysis and conditioning the active ion-exchanger in these membranes
will be as shown in ($\underline{1}$) and ($\underline{2}$), respectively.

(1)   (2)

Electrodes formed by use of TUP gave a Nernstian calibration slope
from 10^{-1} to 10^{-5} M Ca^{2+} and the selectivity of this electrode was
in the order $Ca^{2+} > Mg^{2+} > K^+ > Na^+$, and these data are shown in
Fig. 7. When DAPP was incorporated into the membrane similar
results were obtained (Fig. 8). These changes in the steric and
electronic environment of the sensor group have had little effect
on the selectivity of the electrodes when compared to those prepared using TAP.

In liquid ion-exchanger electrodes it has been recognised that
the solvent-mediator performs an essential function in controlling
selectivity. For example, with calcium bis[di(2-ethylhexyl)phos-
phate] as sensor, a Ca^{2+} ISE results if DOPP is the solvent and a
divalent (water hardness) Ca^{2+}/Mg^{2+} test electrode if 1-decanol is
used.[7] In order that we might introduce a mediator (DOPP) into
our electrodes during the casting process, it was necessary that
we remove the need for a hydrolysis step, since DOPP would also
react with alkali. We have achieved this by incorporating diallyl

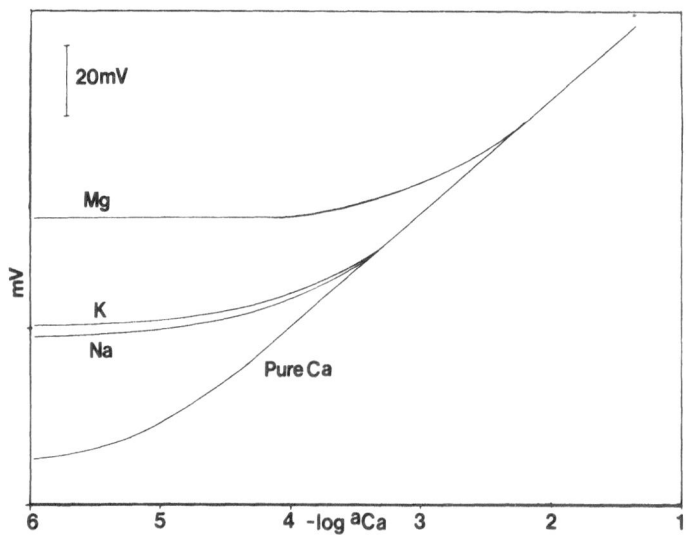

Fig. 7.   Calibration graph for TUP electrode; k_{CaM}^{pot}: Mg, 1.0;
K, 110; Na, 90 at interferent level 10^{-3} M.

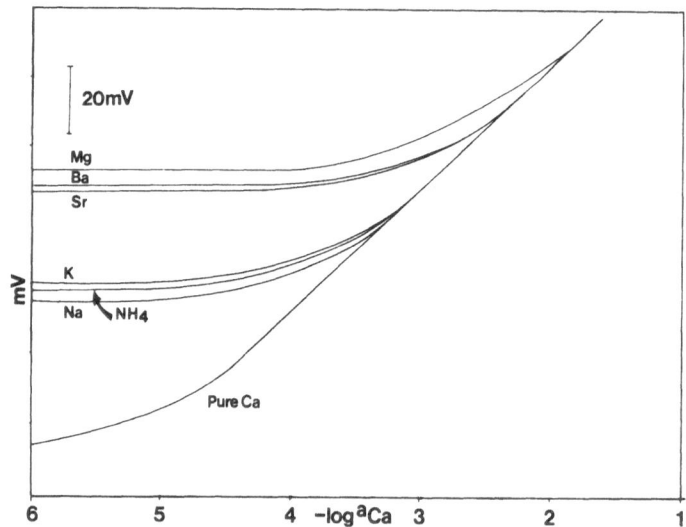

Fig. 8.   Calibration graph for DAPP electrode; k_{CaM}^{pot}: Mg, 2.1;
Ba, 1.05; Sr, 1.0; K, 200; NH_4, 170; Na, 130 at interferent level 10^{-3} M.

phosphoric acid (DAPH) into a membrane. Thus we now have two
routes to a TAP electrode and these are shown in Fig. 9. The
electrochemical response of an electrode made from a membrane
containing 5% DAPH is shown in Fig. 10. A Nernstian calibration
slope was found and the linear response range was similar to that
of the original TAP electrode. These electrodes showed excellent
run-to-run precision and had long lifetimes. However, the selectivity was poorer than that seen with the original TAP electrodes
particularly in the presence of Na^+ ions.

Introducing 10% DOPP as solvent-mediator into a membrane
with 5% DAPH yielded the calcium response and selectivity plots
shown in Fig. 11. The calibration slope was near-Nernstian with
a limit of detection around 10^{-6} M Ca^{2+}. Selectivity coefficients
indicated a distinct improvement in selectivity when compared to a
membrane using 5% DAPH alone, particularly over Mg^{2+}. Fig. 12
shows the response of an electrode fabricated with 25% DOPP together with 5% DAPH. This gave a slightly sub-Nernstian slope
with a limit of detection in the region of 10^{-6} M Ca^{2+}. The
selectivity of this electrode showed a marked improvement over
previous covalently-bound systems and was similar to a PVC electrode (Table 3). However, the lifetime of this system was not
good and the selectivity gradually deteriorated to that of an
electrode containing DAPH alone, indicating that loss of DOPP was
occurring.

In summary, calcium ISE have been produced in which the
sensor groups are covalently-bound. Such electrodes have advantageous properties over available systems in terms of operational
lifetime, response time and ability to withstand hostile environments. The selectivity of electrodes produced using TAP was

Table 3. Selectivity coefficients of membranes[a]

Membrane	k_{CaM}^{pot}	
	Mg^{2+}	Na^+
TAP	0.3	30
DAPH	1.2	250
DAPH/DOPP (1:2)	0.44	200
DAPH/DOPP (1:5)	0.085	8.5
PVC[b]	0.086	1.1[c]

[a] Interferent level 10^{-3} M

[b] Ref. 8

[c] Interferent level 10^{-2} M

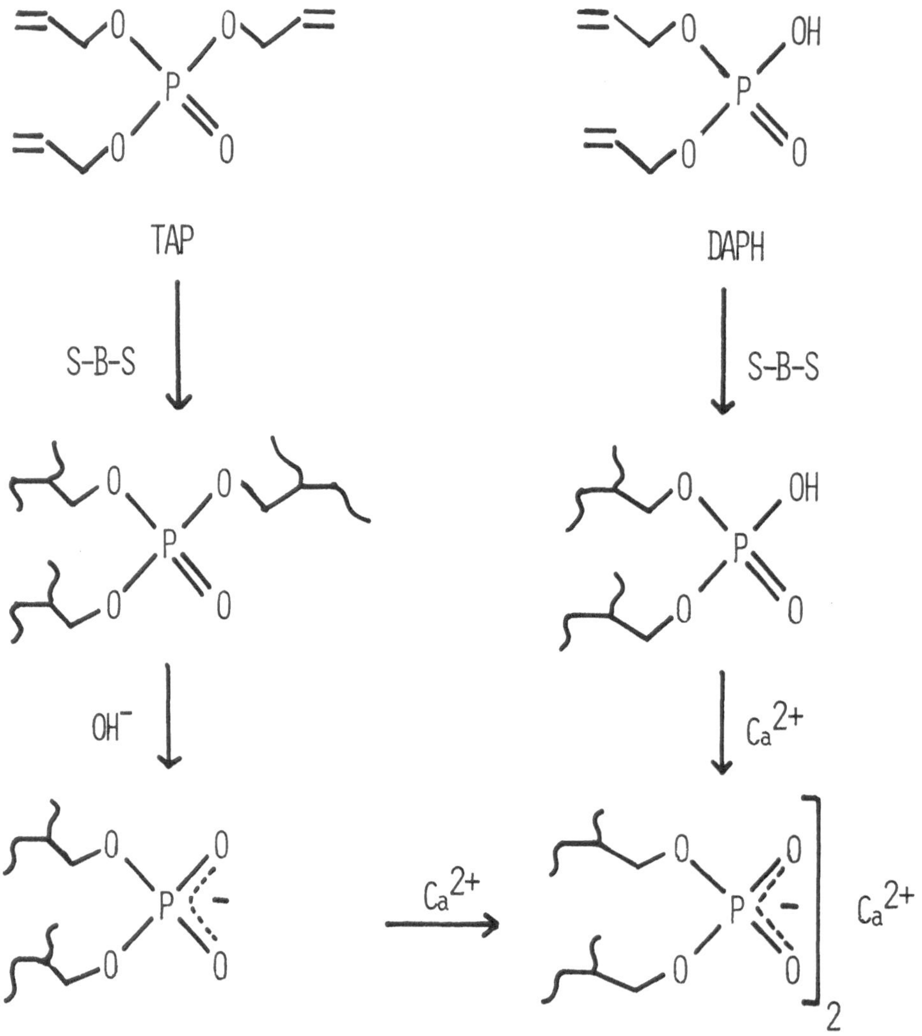

Fig. 9.   Two routes to a covalently-bound dialkylphosphate
          sensor.

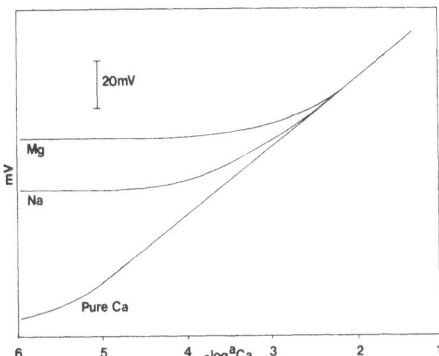

Fig. 10.   Calibration graph for electrode containing 5% DAPH:
k_{CaM}^{pot}: Mg, 1.2; Na, 235 at interferent level 10⁻3 M.

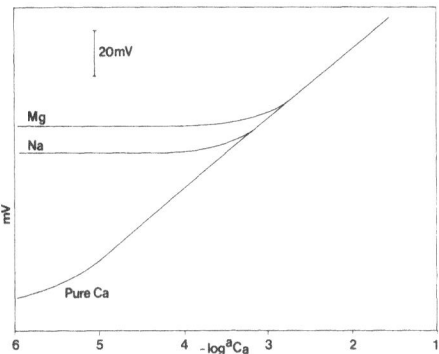

Fig. 11.   Calibration graph for electrode containing 5% DAPH and
10% DOPP; k_{CaM}^{pot}: Mg, 0.44; Na, 200 at interferent level
10⁻3 M.

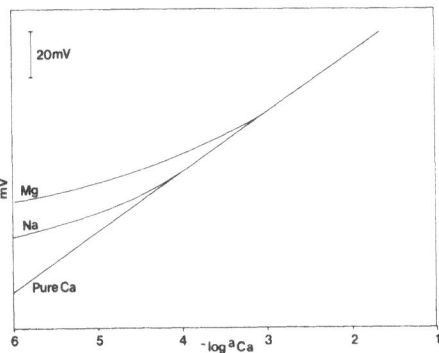

Fig. 12.   Calibration graph for electrode containing 5% DAPH and
25% DOPP; k_{CaM}^{pot}: Mg, 0.085; Na, 8.5 at interferent
level 10⁻3 M.

poorer than for available systems and was not improved by changing the chemical structure of the sensor group. The incorporation of a solvent-mediator, by entanglement, in the membrane has given an electrode with a highly-selective response to calcium over magnesium and sodium ions. Further studies on the production of calcium ISE with covalently bound sensor and mediator groups and the mechanism of their operation are in progress.

REFERENCES

1. J.W. Ross, Science, 156, 1378 (1967).

2. G.J. Moody, R.B. Oke and J.D.R. Thomas, Analyst, 95, 910 (1970).

3. L. Ebdon, A.T. Ellis and G.C. Corfield, Analyst, 104, 730 (1979).

4. A.T. Ellis, G.C. Corfield and L. Ebdon, Anal. Proc., 17, 48 (1980); G.C. Corfield, L. Ebdon and A.T. Ellis, Anal. Proc., in press.

5. Brit. Pat. Appl. 7908452; US Pat. Appl. 8007715.

6. A. Craggs, G.J. Moody and J.D.R. Thomas, J. Chem. Educ., 51, 541 (1974).

7. A. Craggs, L. Keil, G.J. Moody and J.D.R. Thomas, Talanta, 22, 907 (1975).

8. L. Keil, G.J. Moody and J.D.R. Thomas, Analyst, 102, 274 (1977).

ACKNOWLEDGEMENT

     This work was supported by the award of a Society for Analytical Chemistry Studentship (to ATE) from the Analytical Chemistry Trust Fund of the Royal Society of Chemistry.

CHARACTERIZATION OF POLYACRYLAMIDE GELS BY DYNAMIC LIGHT

SCATTERING UNDER THE MICROSCOPE

Nurit Weiss

Polymer Department
Weizmann Institute of Science
Rehovot, Israel

INTRODUCTION

Tanaka, Hocker and Benedek[1] were the first to study dynamic light scattering from gels. Since then several investigations on acrylamide gels utilizing the photon correlation technique were reported.[2-6] Tanaka et al.[2] discovered the existence of a gel spinodal temperature at which the intensity of light scattered by the gel diverges. Hecht et al.[4] found for their gel samples that the elastic modulus, measured from the intensity of quasi-elastically scattered light is in acceptable agreement with the scaling prediction for a good solvent. On the other hand, Gelman et al.[6] found that for their gel preparations (water as solvent in both investigations) the rigidity data follow the scaling law for poor solvents. This indicates that the reaction parameters have an important role on the final structure of the gel preparations.

In this paper the study of laser light scattering from acrylamide gels using an optical microscope is reported. Introducing the microscope is a step towards developing a tool for sub-microscopic probing of network topology. Acrylamide gels are prepared by radical copolymerization which generally leads to network with a spatially inhomogeneous structure. By developing a sensitive permeability apparatus[7] we were able to measure concentration distribution characterizing the gels while the friction coefficient f between the network segments and the solvent turned out to be essentially constant. It was found that the permeability of acrylamide gels increases strongly with increasing degree of

crosslinking, indicating extensive non-random crosslinking. This
non-random crosslinking may result in poor mechanical performance[8,9]
of the inhomogeneous network. Through viscoelastic measurements
it was indeed shown that only a small fraction of the crosslinks
in acrylamide gels is elastically effective. These measurements
induced our suggestion[10] that depending on the reaction parameters
active centers in the gel reaction mixture are formed connected
to each other by much more diluted zones. Each zone should be
characterized by its own permeability and elastic modulus governing
its equation of motion, and thus its light scattering properties.
These zones are expected to contribute differently to the overall
observed time correlation function of the intensity of the scattered light. The resolving power of the microscope is used in this
work to detect the existence of different zones in the gels by
decreasing the dimensions of the scattering volume down to regions
of the order of magnitude of the wavelength of the light.

EXPERIMENTAL

A. Materials and Sample Preparation

Polystyrene latex (Dow Chemicals, Diameter = .088µ, St. Dev.
= .008µ) was used in dilute suspension in water to establish the
feasibility of laser light scattering measurements from extremely
small regions.

Acrylamide Gels

These were prepared as before.[8] The gelation mixture was
filtered (Millipore .22µ Gs) into a special square cuvette.
One face of this cuvette was replaced by an ultrathin cover-slide
to allow the highest magnification objective lens used to approach
the gel at its proper working distance. Seven different transparent gels of total concentration c = 4,7,9,10,12,14,16 g/dl
were examined under the microscope. The crosslinker concentration
was 2 percent of the total monomer concentration.

B. Methods

The set-up: (see Fig. 1). The sample was placed on the microscope stage (Zeiss research microscope) and was illuminated by a
small pencil of light (sharply focused ∿50µ diameter) from a
40 MW He-Ne laser beam. The microscope objective collected the
scattered light in a median direction of 90°. Three different
objective lenses (Zeiss) were used x 10 N.A. = .25, x 40 N.A. = 40,
x 100 N.A. = 1.30 (oil). The scattered light was transmitted into
a photomultiplier tube (ITT F.W. 130) by an optical fiber light
guide (50µ diameter; Gamma Sci. Inc., Model 700-10-36A) whose

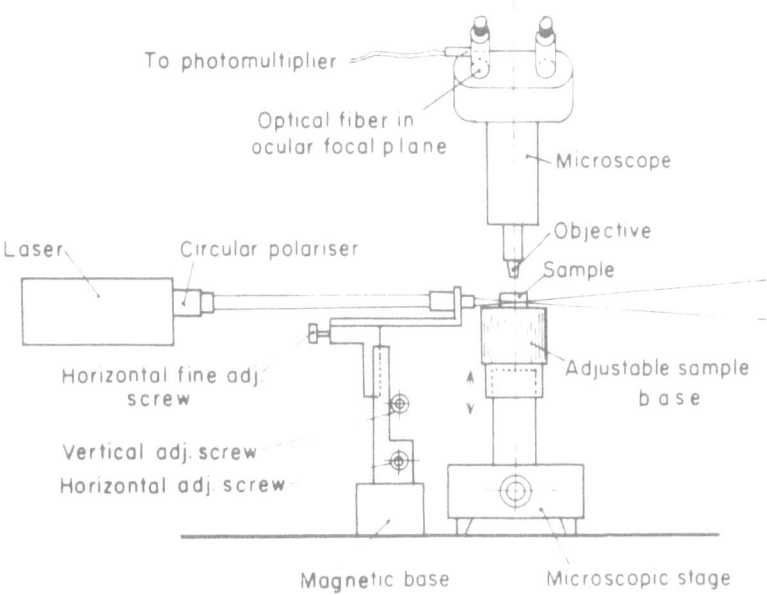

To photomultiplier

Optical fiber in
ocular focal plane

Microscope

Objective

Laser Circular polariser

Sample

Horizontal fine adj.
screw

Adjustable sample
base

Vertical adj. screw
Horizontal adj. screw

Magnetic base Microscopic stage

Fig. 1: Set-up for laser light scattering under the microscope.

receiving end is centered in the first focal plane of the ocular
lens. A 24-channel digital correlator (Malvern type 4300) was
used to obtain the photocurrent correlation function. The photo-
count clipped correlation functions were normalized against the
background by the available instrument normalization procedure.
Ten different clock settings were collected in order to cover a
wide range of correlation times. The analysis of the functions
was performed according to the Koppel cumulants method as extended by Mazer[11]. Assuming a homodyne beating experiment this
analysis was conducted on the combined ten normalized functions
obtained which together formed a smooth line (see Fig. 2).

The experiments were conducted as follows: First the time
autocorrelation function was obtained through the x10 objective

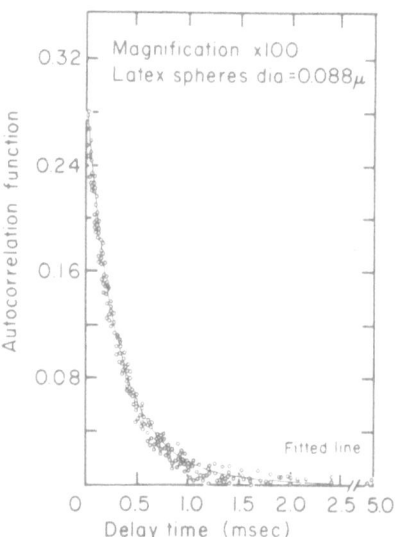

Fig. 2: Time autocorrelation function of intensity of light
scattered from latex suspension in the highest magnification used.

lens. Enough data was collected and the objective lens was then
carefully replaced by the lens of next higher magnification and
so on: This was done taking great care not to move the gel sample
in the laser beam.

RESULTS

Calibration with Latex

     Typical correlation function obtained from the scattering of
suspensions of .088μ diameter polystyrene latex spheres when imaged
to the optical fiber with the x100 magnification objective lens is
shown in Fig. 2. The three objective lenses gave the same results.

The correlation function was not purely exponential but the mean
decay rate calculated was the same for the three objectives and
corresponded to a mean diameter of the sphere of .0874 ± .003μ
given by the manufacturer. When attempting to fit the correlation
function by a two term representation (as was suggested by Mazer)
the calculated maximum diameter of the sphere never exceeded that
of the Millipore filter pore diameter through which the latex
sample has been passed.   This two component fit is also given
in Fig. 2.

Scattering from Gels

In contrast to the work with the latex suspensions, the decay
ratio of the normalized time autocorrelation function of light
scattered from gel samples did depend on the magnification of the
collecting objective lens. In the case of x100 objective lens,
the ten autocorrelation functions (for the ten different clock
settings) could not be reduced to a single master curve even after
performing the normalization procedures, though each function looked
exponential. A smooth curve could be derived from the normalized
and then combined correlation functions in the case of x10 and x 40
magnification collecting objective lenses. The decay rates and the
(collective diffusion constants, D_c, derived from them)are strong
functions of the magnification (see Fig. 3). Interestingly enough,
there is a tendency for these to agree in the case of the 4 and
16% gels, i.e. at the extremes of the concentration range tested.

DISCUSSION

The principal cause for light scattering is the local fluctuation in dielectric constant which arises from concentration fluctuations. These concentration fluctuations may be described
through the vector displacement u defined as the vector connecting
the positions of a point (in our case, in the network) before and
after a change in its place has occurred.[12]   The displacement u is
thus a function of position and time. Tanaka et al.,[1] suggested
to modify the equation of motion for an elastic medium[12]

$$\rho \ddot{u}_i = \partial \sigma_{ij}/\partial x_j \tag{1}$$

(ρ - density of the medium, $\ddot{u}$ - acceleration, $\partial \sigma_{ij}/\partial x_j$ - internal
stress force) by adding the drag caused by friction, between the
gel and the liquid, to the right hand side of eq.(1). They thus
obtained the equation of motion for a viscoelastic medium

$$\rho \ddot{u}_i = \partial \sigma_{ij}/\partial x_j + f\dot{u}_i \tag{2}$$

($\dot{u}_i$ - velocity, f - specific resistance to flow).

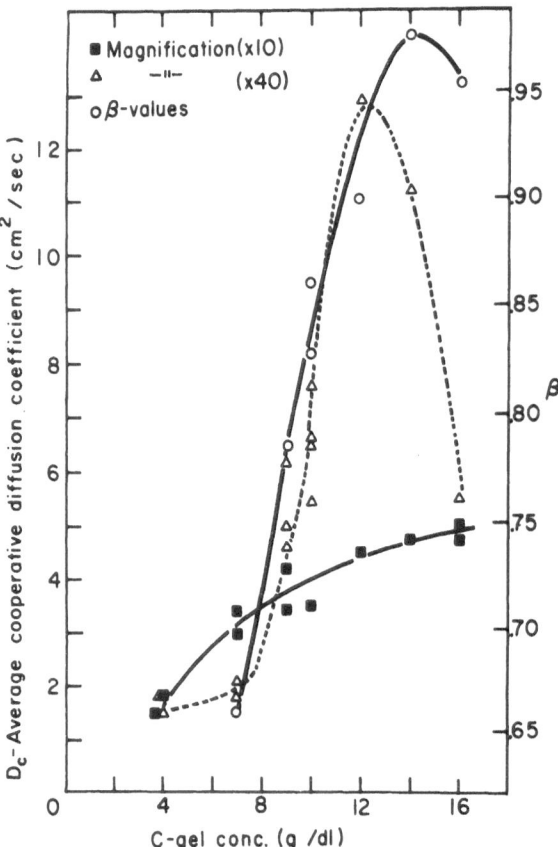

Fig. 3: Dependence of D_C on gel concentration as calculated from decay rates obtained using x10 and x40 magnification objective lenses. β function explained in text.

Expressing the stress tension σ_{ij} in terms of derivatives (with respect to position) of $\underline{u}$ and the bulk and shear modulus K and μ of the network respectively, they derived the time correlation function of the displacement $\underline{u}$. This is directly connected to the time correlation function of the intensity of the scattered light. Thus, they showed that the decay rate Γ of the time auto-correlation function $R(\tau)^{1/2}$ is given by

$$\Gamma = D_c q^2 \tag{3}$$

where D_c is the cooperative diffusion coefficient of the network given by

$$D_c = 2\mu/f \tag{4}$$

and q is the wave vector of the scattering fluctuations.

In heterogeneous gel structures the correlation function can be expected to contain a variety of components characterized by different decay constants $\{\Gamma_i\}$ $R(\tau)^{1/2}$ has to be generalized to a weighted sum of decaying exponentials

$$R(\tau)^{1/2} = \sum_{i=1}^{\infty} G_i e^{-\Gamma_i \tau} \qquad \text{with} \sum_{i=1}^{\infty} G_i = 1 \tag{5}$$

The possibility to isolate these components by using the microscope is now discussed. First, it may be concluded from Fig.2 that although the measured $R(\tau)^{1/2}$ functions were not pure exponents they correlated at all magnification used for the latex particle scatterers. According to the Rayleigh criterion the linear separation of two just resolvable point objects is inversely proportional to the numerical aperture of the focusing lens. This might be in conflict with the requirement of a well defined wave vector q of the scattering fluctuations (due to the wider acceptance angle). Fig. 2 shows that even for the x100 magnification objective lens, the wave vector q is still acceptably defined for our latex systems. The results for the gel preparations, however, seem to reflect their special structure in contrast to homogeneous solutions.

The fact that no master curve could be obtained when the gel samples were imaged to the optical fiber with the x100 magnification lens, may give some indication of the size of the regions which scatter differently in the gel. If the number of scatterers in the imaged volume is very low because their size tends to the ratio of optical fiber diameter to linear magnification ($50\mu/100$ in our case) the number of scatterers getting in and out the investigated volume element will fluctuate violently and no single master curve will be achieved. The size of the heterogeneous regions in our suggested model[10] agree quite well with these results. Increasing the number of crosslinks, reduces the size, thus improving the possibility of scanning these gel preparations with the x100 magnification objective. This has not yet been checked.

The dependence of the D_c values on gel concentration is given in Fig. 3. For the x10 magnification lens the curve drawn fits a power law $D_c \sim c^m$ with M = .88. For semi-dilute solutions in a poor solvent, where strong interactions between overlapping polymer chains are to be expected, D_c is predicted to be a linear function of the concentration, while in good solvent m is predicted to be .75.[13]

The results for the x40 magnification lens may be explained by assuming that the contribution to scattering from the dilute gel regions can be neglected when the scattering volume has been decreased by the use of the x40 magnification lens. The more dense gel regions, the only ones now seen have higher D_c values and thus the average is increased considerably. In our model for heterogeneous gels we defined a parameter β which is the volume fraction of the dense region. β is thus a measure of the probability to encounter dense regions. We found that for the gel systems used for the present investigation, β increases with the gel concentration up to a maximum then falls off. The o-heavy line in Fig. 3 is the plot of β against c which was obtained.[10] Its shape is in remarkable agreement with the shape of the D_c plot for the x40 magnification lens.

In our model[10], we also calculated all the necessary parameters to evaluate the cooperative diffusion coefficient of the slower (dilute) phase $D_{c,dil}$. By assuming

$$\mu = \nu kT \qquad\qquad (6)$$

where ν is the number density of crosslinks in the dilute phase and

$$f = \frac{\eta}{K_s} \qquad\qquad (7)$$

η - solvent viscosity, K_s - permeability coefficient of the dilute phase, we may calculate $D_{c,dil}$ and compare it to the values obtained by the two component fit (which gives the diffusion coefficients of the fastest $D_{c,f}$ and slowest $D_{c,s}$ components in the scattering volume). This comparison could be done only, of course, for the results with the x10 magnification lens. Fig. 4 shows the fitted $D_{c,s}$ and calculated $D_{c,dil}$ values as a function of c. At low concentration, there is a remarkable agreement but above c = 10 g/dl the results diverge. An explanation might be advanced following Hecht et al.[4]. The $D_{c,s}$ values calculated from the correlation function reflect the high frequency response of the dilute region to deformation. Here entanglements might contribute to the elasticity of the network. (The deformations and the deformation rates are of course synomymous with the concentration fluctuation and their time scale) The calculated $D_{c,dil}$ values are related to the permanent crosslink density. As the overall concentration of gel sample increases the concentration of the dilute region increases[10] and the effect of entanglements increases in accordance.

REFERENCES

1. T. Tanaka, L.O. Hocker, and G.B. Benedek,J.Chem.Phys.59:5151(1973)
2. T. Tanaka, S. Ishiwata, and C. Ishimoto, Phys.Rev.Lett.

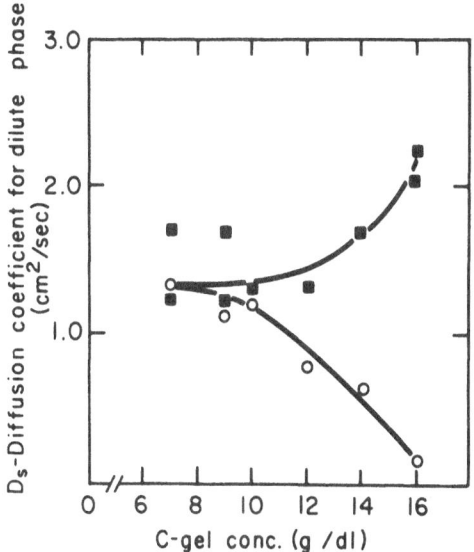

Fig. 4:   Dependence of $D_{c,dil}$, o and $D_{c,s}$ ■ on gel concentration.

          38:771 (1977).
3.   T. Tanaka, Phys.Rev.A, 17:763 (1978).
4.   A.M. Hecht and E. Geissler, J.Physique 39:631 (1978).
5.   T. Tanaka, Polymer 20:1404 (1979).
6.   R.A. Gelman and R. Nossal, Macromolecules 12:311 (1979).
7.   N. Weiss and A. Silberberg in "Hydrogels for Medical and Related
        Applications" Ed. J.O. Andrade, A.C.S. Symposium Series 31,
        Washington, D.C. (1976) p.69.
8.   N. Weiss and A. Silberberg, The British Polymer J. 9:144 (1977).
9.   V.F. Janas, F. Rodriguez, and C. Cohen, Private Communication

10. N. Weiss, T. van Vliet, and A. Silberberg, J.Pol.Sci.
       17:2229 (1979).
11. D.E. Koppel, J.Chem.Phys. 57:4814 (1972).
12. L.D. Landau and E.M. Lifshitz, in "Theory of Elasticity"
       Sec.Eng.Ed., Pergamon Press (1970).
13. F. Brochard and P.G. deGennes, Macromolecules 10:1157 (1972).

THERMAL CROSSLINKING OF A

CHEMICALLY-MODIFIED IONOMER

M. J. Covitch, S. R. Lowry,*
C. L. Gray, and B. Blackford

T. R. Evans Research Center
Diamond Shamrock Corporation
Painesville, Ohio 44077

INTRODUCTION

Perfluorosulfonic acid ionomer resins have received considerable attention by the chlor-alkali industry as chemically inert polymers suitable for use as cation exchange membranes in sodium chloride electrolysis cells. A paper by Hora and Maloney[1] reported that the low cation selectivity of standard perfluorosulfonic acid membranes is improved by amine modification of the sulfonyl halide form with ethylenediamine (EDA). Several related patents[2] describe this reaction and its consequences on membrane selectivity. In addition to direct amidation, the patents refer to the process of post heat treatment which results in crosslinking by the difunctional amine. This paper summarizes an investigation of the chemistry of this thermal crosslinking reaction.

EXPERIMENTAL

0.18 mm thick sheets of Nafion[3] 117 perfluorosulfonic acid membranes (0.91 meq/g dry resin) were converted to the reactive sulfonyl chloride form by refluxing in a 33 weight % solution of PCl_5 in $POCl_3$ for 96 hours. Excess PCl_5 and $POCl_3$ was removed by boiling in CCl_4 for several hours. After vacuum drying at 80°C for 24 hours, the sulfonyl chloride intermediate was converted to the sulfonamide form by contacting the polymer with a 95% ethylene-

*Present address: Nicolet Instrument Corporation, Madison, WI 53711

diamine solution (5% water) at room temperature for up to 250 hours. Excess EDA was extracted with alternate acetic acid and water rinses.

The stoichiometry of the EDA/polymer reaction was studied by suspending pre-weighed, dried films in 95% EDA for various amounts of time, extracting excess EDA, and drying to constant weight under vacuum at 80°C. The EDA penetration depth was measured on thin microtomed cross-sections by optical microscopy.

Thermal gravimetric analysis (TGA) was performed on freely-hanging films in a Perkin-Elmer TGS-2 at a scanning rate of 5°C/min in dry air.

A Nicolet 7199 Fourier Transform Infrared Spectrometer, equipped with a glass pyrolysis gas cell (10 cm path length), was used to detect gases which were evolved as the sample was heated.

Dynamic mechanical spectra were obtained at 110 HZ on a Rheovibron DVD-II Viscoelastometer with the sample chamber flooded with dry nitrogen.

RESULTS AND DISCUSSION

Nafion[3] perfluorosulfonic acid resin in the sulfonyl chloride form is essentially a copolymer of tetrafluoroethylene and per-fluoro(3,6-dioxa-4-methyl-7-octenesulfonyl chloride):

$$\left(CF_2 - CF_2\right)_n \quad \left(CF_2 - CF\right)_m$$
$$| $$
$$O-CF_2-CF-O-CF_2-CF_2-SO_2Cl$$
$$|$$
$$CF_3$$

The quantitative conversion of the initial sulfonic acid polymer to its sulfonyl chloride form was confirmed by the disappearance of the S-O symmetric stretch ($1060 cm^{-1}$) infrared absorbance assigned to the sulfonic acid moiety and the appearance of a strong S-O asymmetric stretch band at $1425 cm^{-1}$ due to the sulfonyl chloride group.

The initial reaction of ethylenediamine with the sulfonyl chloride intermediate may proceed according to either of the following possible reactions:

$$RSO_2Cl+EDA \rightarrow RSO_2NHCH_2CH_2NH_2 + HCl \qquad (1)$$

or

$$2RSO_2Cl+EDA \rightarrow RSO_2NHCH_2CH_2NHSO_2R + 2HCl \qquad (2)$$

Let W_1 be defined as the theoretical change in membrane weight, assuming total conversion, according to reaction (1). W_2 is then the theoretical change in weight according to reaction (2).

$$W_1 = \frac{Wi}{1118} \ (59-35.4)$$

$$W_2 = \frac{Wi}{1118} \ (29-35.4)$$

where Wi = initial dry weight of the sulfonyl chloride film.

The observed change in membrane weight, ΔW, is plotted versus EDA treatment time in figure 1. This stoichiometric data strongly supports the predominance of reaction (1), i.e. the unifunctional reaction of ethylenediamine with the sulfonyl chloride group, since the ratio of $\Delta W/W_1$ approaches unity asymptotically with time. Total nitrogen analysis confirms this conclusion (observed: 2.3%N; calculated(1): 2.5%N; calculated(2): 1.2%N). Reaction(2) is not favored because the average distance between sulfonyl chloride groups (assuming random spacial distribution) in the polymer is nearly three times the end-to-end distance of the EDA molecule. Gravimetric techniques are not sufficiently sensitive to rule out a small percentage of EDA crosslinks, however.

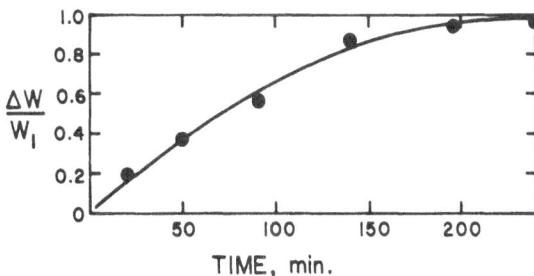

Fig. 1.   Observed (ΔW) relative to calculated (W_1) weight change upon EDA reaction as a function of treatment time, assuming unimolecular reaction.

Since the EDA reaction is diffusion-controlled, a sharply-defined reaction front is observed as the reaction proceeds toward the interior of the membrane (see figure 2). When the depth of penetration (d/2) is compared to the stoichiometric data ($\Delta W/W_1$), it can be demonstrated that reaction is complete in this reaction zone except within approximately 0.009 mm of the reaction front. In figure 3 is plotted the fractional EDA treatment depth (d/to, where to = membrane thickness) versus the volume fraction of membrane converted to the sulfonamide ($\Delta W/W_1$). If the observed treatment zone was fully reacted, the data would fall along the dotted line in figure 3. The data are displaced along a parallel line, however, indicating that as the reaction proceeds toward the interior, a well-defined sulfonamide concentration profile exists at the reaction front as shown schematically in figure 4. As the treatment layer (d/2to) increases in thickness, the frontal zone thickness remains constant leaving a fully reacted region of membrane behind. Since the large proportion of sulfonyl chloride groups in the reacted layer are indeed tagged with an EDA molecule, the predominant post-crosslinking reaction within the EDA layer cannot involve reaction of the free amino group with an unreacted adjacent sulfonyl chloride group.

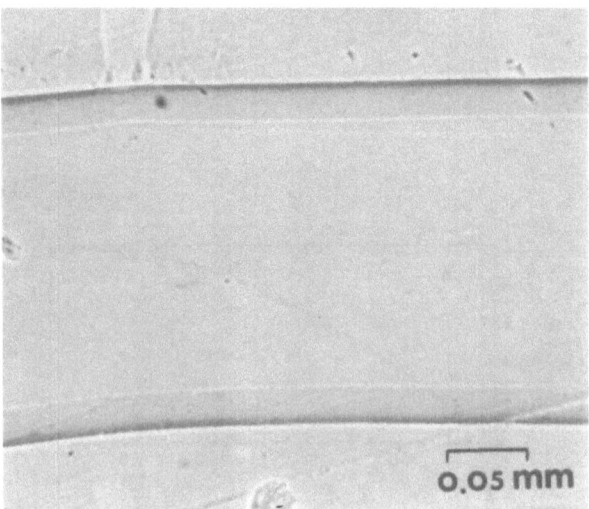

0.05 mm

Fig. 2. Optical micrograph of a partially EDA-modified membrane showing the sharp boundary between the sulfonate and sulfonamide regions.

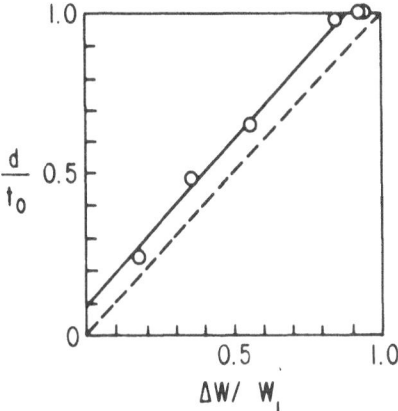

Fig. 3.   Fractional EDA treatment depth (d/to) vs. volume fraction
          of EDA-reacted material ($\Delta W/W_1$).

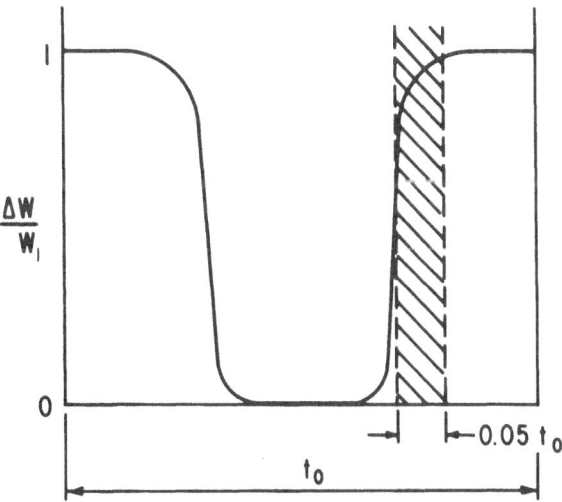

Fig. 4.   Schematic representation of the sulfonamide concentration
          profile across the thickness of the membrane. Shaded area
          represents the incompletely-reacted frontal zone.

When an EDA-modified film was heated within a nitrogen-purged
infrared gas pyrolysis cell, ammonia was liberated (see figure 5)
above 90°C at a rate which increased with temperature (figure 6).
Some EDA was also observed in the vapor space, although it is
considered to have been residual trapped EDA since its rate of
evolution increased 8-fold above its boiling point. The free
amino group of the sulfonamide sidechain is the most probable
source of ammonia according to several possible reactions:

$$RSO_2NHCH=CH_2 \qquad\qquad (3)$$

$$RSO_2NHCH_2CH_2$$
$$\qquad\qquad\qquad\searrow$$
$$\qquad\qquad\qquad\qquad NH \qquad\qquad (4)$$
$$\qquad\qquad\qquad\nearrow$$
$$RSO_2NHCH_2CH_2$$

$$RSO_2N\underset{CH_2}{\overset{CH_2}{\diagdown\;|\;\diagup}} \qquad\qquad (5)$$

$$RSO_2N\underset{CH_2CH_2}{\overset{CH_2CH_2}{\diagdown\quad\diagup}}NSO_2R \qquad\qquad (6)$$

The infrared spectra of films heated at 115°C for 24 hours show no
evidence of vinyl, aziridine, or piperazine structure, although there
is some decrease in the absorbance intensity in the NH stretch region.
Thus, the diethylenetriamine crosslink (4) is suggested.

Further support for this reaction is provided by measurements
of sample weight loss upon heating. Reactions (3), (5), and (6)
lose one mole of NH_3 per equivalent whereas reaction (4) loses one
mole of NH_3 for every two equivalents. The observed isothermal
weight loss in the 100-130°C region (0.47%) is within reasonable
agreement with that calculated for reaction (4) (0.42%), whereas
0.84% of the initial sample weight would have been lost according
to the other reaction schemes.

Before postulating a chemical mechanism for this crosslinking
reaction, several examples of the effects of post-heat treatment
on physical properties will be mentioned. The thermal stability
of EDA-modified films has been studied by thermal gravimetric
analysis (figure 7) relative to a standard potassium sulfonate
membrane and an ethylamine-modified sample.

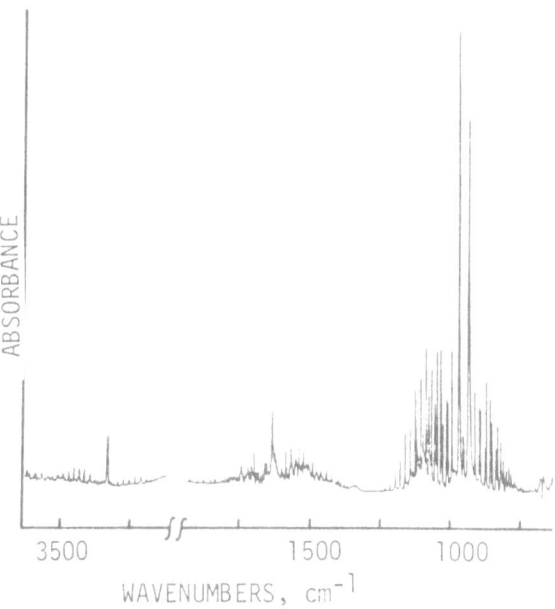

Fig. 5.   Infrared absorbance spectrum of the gas space above an EDA-
          modified membrane at 130°C. Spectrum matches that of
          gaseous ammonia.

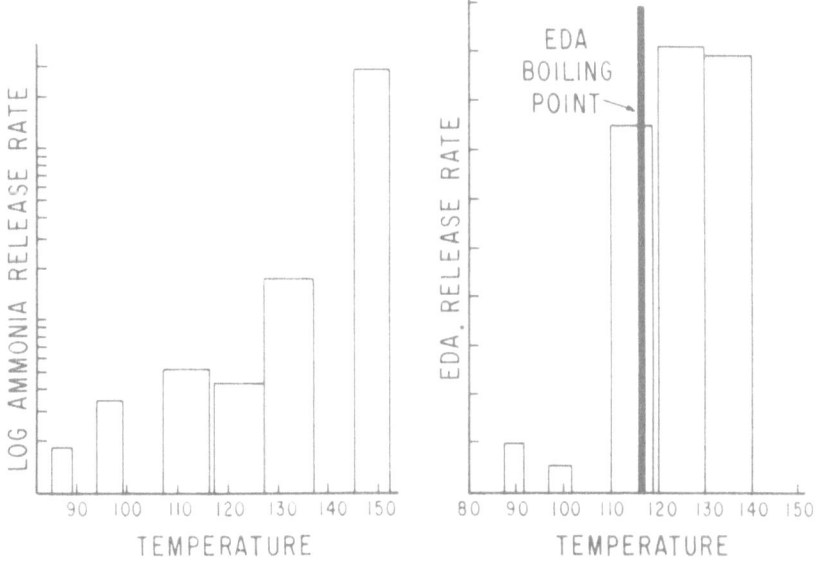

Fig. 6.   Gas release rates (arbitrary units) of ammonia and EDA at
          various temperatures (°C). Infrared absorbance peak
          intensities at 970 cm^{-1} and 770 cm^{-1} were used to follow
          the buildup of ammonia and EDA respectively.

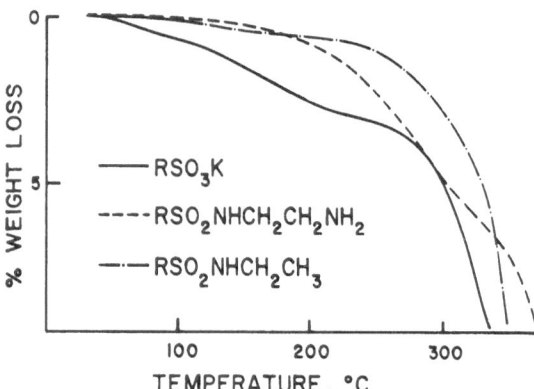

Fig. 7. Thermal gravimetric analyses of membranes in the potassium
         sulfonate, EDA sulfonamide, and ethylamine sulfonamide
         forms.

The potassium sulfonate displays a sharp weight loss above 300°C
which is mirrored (perhaps at lower temperatures) by the thermal
behaviors of both sulfonamide membranes. However, above 300°C, the
weight loss of the EDA membrane is inhibited for nearly 50 degrees.
This inhibition is not observed for ethylamine samples, suggesting
that molecular bonding (leading to enhanced thermal stability) takes
place as a direct consequence of the free amino group.

As expected, crosslinking reduces the degree of membrane
swelling in compatible liquids. The weight of an as-reacted EDA
membrane gained 8.3% when immersed in 10$\underline{M}$ NaOH at 80°C. After
heating at 115°C for 24 hours, the membrane weight gain under the
same conditions dropped to 6.3%.

The effect of sulfonamide crosslinking on the dynamic modulus
of an EDA-modified film is shown in figure 8. The sample was
heated from room temperature to 180°C at a rate of 1°C/min. during
the first run and rapidly cooled.

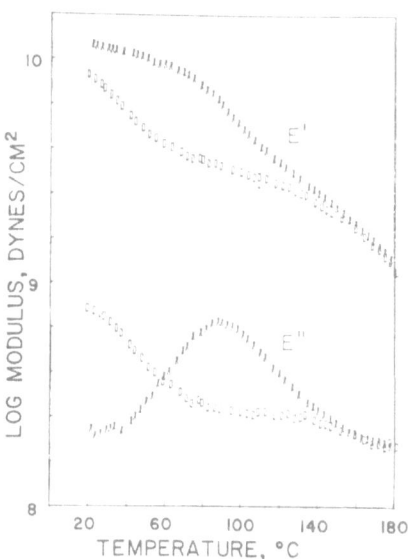

Fig. 8.   Storage (E') and loss (E") moduli as functions of tempera-
          ture at 110 HZ for an EDA-modified membrane.   (▯) data
          from the first heating; (X) data from the second heating.

Structural changes which occurred during the first heating were
manifested in an elevated value of the storage modulus (E') and an
increase in the E" peak temperature (nearly 70°C) for the second
run. The E" peak is related to relaxations which can be associated
with the sulfonamide group since (1) the peak intensity is proportional to the sulfonamide content, and (2) the standard perfluorosulfonate ionomer[4] does not exhibit a relaxation in this

region. Thus, the crosslinks are observed to raise the activation
barrier for sulfonamide relaxation by imposing the constraint of
cooperativity on the motions of each group.

Ammonia evolution from thermally-degraded amides and sulfonamides has been previously reported.[5-7] The imidization of polyacrylamides[5,6] involves the formation of a six-member ring with
elimination of ammonia:

$$-CH_2-CH-CH_2-CH-CH_2- \quad \xrightarrow{\Delta} \quad -CH_2-CH-CH_2-CH-CH_2- \quad +NH_3$$

This ring closing reaction is aided by the close proximity of
neighboring amides. Analogously, the sulfonamide groups in per-
fluorosulfonamide ionomers are presumably segregated into domains
(similar to ion clusters in perfluorosulfonate polymers)[8] which
promotes close contact between neighboring EDA sulfonamides:

$$- OCF_2CF_2 - SO_2NHCH_2CH_2NH_2$$

$$- OCF_2CF_2 - SO_2NHCH_2CH_2NH_2$$

With the sidechains relatively fixed at the perimeter of each
domain (denoted by the dashed line above), the sulfonamide cross-
linking reaction (4) can take place in a pseudo ring closing fashion.

The mechanism of crosslinking probably involves an aziridine
synthesis followed by ring opening. The cyclization of sulfonamides[9] is a common example of aziridine ring formation by intra-
molecular displacement by the amide ion.

L= leaving group
A= acyl group

The terminal amino group of the EDA sulfonamide is a good leaving
group, and the zwitterionic nature of this structure promotes
displacement even at neutral pH.

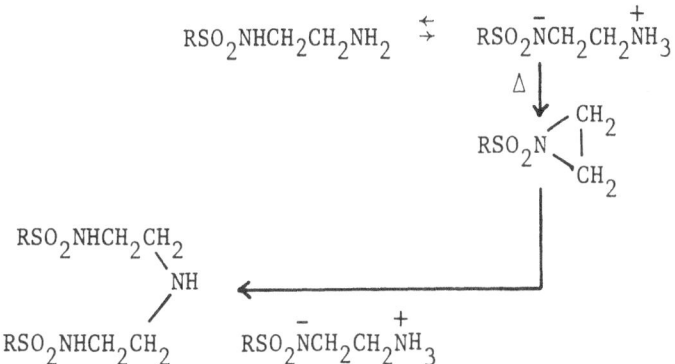

Attack by a neighboring EDA sulfonamide opens the three-membered ring to result in the diethylenetriamine crosslink.

Although direct spectroscopic confirmation of the proposed cross-link structure has not been obtained, future ^{15}N NMR studies on ^{15}N enriched EDA sulfonamide membranes are planned to provide more detailed information.

REFERENCES

1. C. J. Hora and D. E. Maloney, Electrochem.Soc. Extended Abstr., 77(2), 1145 (1977).
2. W. G. Grot, U.S. patents 3,969,285 (1976); 4,026,783 (1977); 4,030,988 (1977). Also S. K. Burkhardt and D. E. Maloney, U.S. patent 4,168,216 (1979).
3. Nafion is a registered trademark of E. I. duPont de Nemours & Co., Inc. for its perfluorosulfonic acid products.
4. S. C. Yeo and A. Eisenberg, J. Appl. Polym. Sci., 21, 875 (1977).
5. J. Zurakowska-Orszagh, W. Busz, and K. Soerjosoeharto, Bull. Acad. Pol. Sci., Ser. Sci. Chim., 25 (11), 845 (1977).
6. I. C. McNeill and M. Zulfiqar, J. Polym. Sci.: Polym. Chem. Ed., 16, 2465 (1978).
7. D. Schittenhelm and K. Hermann, Dtsch. Apoth.-Ztg., 110 (39), 1441 (1970).
8. T. D. Gierke, Electrochem. Soc. Extended Abstr., 77(2), 1139 (1977).
9. O. C. Dermer and G. E. Ham, "Ethyleneimine and Other Aziridines, Chemistry and Applications," Academic Press, New York (1969).

CHARACTERIZATION OF IONICALLY-CROSSLINKED

POLY(ACRYLIC ACID) MEMBRANES

A. C. Habert

COPPE, Fed. University of Rio de Janeiro
C. P. 1191, Rio de Janeiro, Brazil

N. Jarvis, C. M. Burns and R. Y. M. Huang

University of Waterloo
Department of Chemical Engineering
Waterloo, Ontario, Canada N2L3G1

ABSTRACT

Recently-developed ionically-crosslinked polymer membranes were characterized in order to elucidate some of their transport properties. Hydrophilic membranes are obtained by crosslinking polymers containing carboxylic pendent groups with aluminum salts. Infrared spectroscopy indicates the presence of two absorption bands which were attributed to the carboxyl-aluminum structure. Mechanical strength studies show that tensile strength (in the wet state) is a function of metal content, and compares favourably with cellulose acetate. Scanning electron microscopy primarily reveals a "dense" structure. However, modification in synthesis conditions indicates the possibility of structural modifications.

CONTRIBUTORS

Thomas V. Alfredson, Varian Instrument Group, 2700 Mitchell Drive,
    Walnut Creek, CA 94598

Claude Benezra, Laboratoire de Dermato-Chimie Clinique
    Dermatologique, CHR de Strasbourg, Strasbourg, F-67091, France

A. Berg, Department of Chemistry and Chemical Engineering,
    Michigan Technological University, Houghton, MI 49931

B. Blackford, T. R. Evans Research Center, Diamond Shamrock
    Corporation, Plainesville, OH 44077

Brian A. Bolto, CSIRO Division of Chemical Technology, South
    Melbourne 3205, Australia

Robin G. Booth, Dynapol, 1454 Page Mill Road, Palo Alto, CA 94304

I. J. Brass, Chemistry Department, University of Aberdeen,
    Scotland

T. S. Brun, Department of Chemistry and Chemical Engineering,
    Michigan Technological University, Houghton, MI 49931

C. M. Burns, University of Waterloo, Department of Chemical
    Engineering, Waterloo, Ontario, Canada N2L3G1

Eleanor R. Carlson, Amicon Corporation, 25 Hartwell Avenue,
    Lexington, MA 02173

Anthony R. Cooper, Dynapol, 1454 Page Mill Road, Palo Alto, CA
    94304

G. C. Corfield, Department of Chemistry, Sheffield City
    Polytechnic, Sheffield S1 1WB, England

M. J. Covitch, T. R. Evans Research Center, Diamond Shamrock
    Corporation, Painesville, OH 44077

John V. Dawkins, Department of Chemistry, Loughborough University
    of Technology, Loughborough, Leicestershire LE11 3TU, England

Charles W. Desaulniers, Millipore Corporation, Bedford, MA

L. Ebdon, Department of Chemistry, Sheffield City Polytechnic,
    Sheffield S1 1WB, England

A. T. Ellis, Department of Chemistry, Sheffield City Polytechnic,
    Sheffield S1 1WB, England

M. R. Faith, Institute for Cancer and Blood Research,
    Beverly Hills, CA 90211

Jean M. J. Fréchet, Department of Chemistry, University of Ottawa,
    Ottawa, Ontario K1N-9B4, Canada

H. L. Frisch, Department of Chemistry, SUNY at Albany,
    Albany, NY 12222

Scott P. Fulton, Amicon Corporation, 25 Hartwell Avenue,
    Lexington, MA 02173

C. M. Garlock, College of Physicians and Surgeons, Columbia
    University, New York, NY

C. L. Gray, T. R. Evans Research Center, Diamond Shamrock
    Corporation, Painesville, Ohio 44077

A. C. Habert, COPPE, Fed. University of Rio de Janeiro,
    C. P. 1191, Rio de Janeiro, Brazil

Jay M. S. Henis, Corporate Research and Development, Monsanto
    Company, St. Louis, MO 63166

B. H. J. Hofstee, Biochemistry Division, Palo Alto Medical
    Research Foundation, Palo Alto, CA 94301

C. J. Hou, Dept. of Chemistry and Polymer Res. Inst., Polytechnic
    Inst. of N. Y., 333 Jay St., Brooklyn, NY 11201

R. Y. M. Huang, University of Waterloo, Department of Chemical
    Engineering, Waterloo, Ontario, Canada N2L3G1

N. Jarvis, University of Waterloo, Department of Chemical
    Engineering, Waterloo, Ontario, Canada N2L3G1

Nava Kahana, Department of Organic Chemistry, The Weizmann
    Institute of Science, Rehovot, Israel

Kenji Kamide, Textile Research Laboratory, Asahi Chemical
Industry Comp. Ltd., Takatsuki, Osaka, Japan

S. Kato, Department of Physiology, School of Medicine,
Yamaguchi University, U b e, 755, Japan

Shigeo Katoh, Chemical Engineering Department, Kyoto University,
Kyoto 606, Japan

Toru Kawai, Department of Polymer Technology, Tokyo Institute of
Technology, Ookayama, Tokyo, Japan

M. Khojasteh, Dept. of Chemistry and Polymer Res. Inst., Polytechnic Inst. of N.Y., 333 Jay St., Brooklyn, NY 11201

K. Kock, Forschungsinstitut Berghof GmbH, D - 7400 Tübingen,
Postfach 1523

A. E. Kreituss, Institute of Wood Chemistry, Academy of Sciences
of the Latvian SSR, Riga, Latvian SSR 226006, USSR

P. D. Lindley, Hyland Therapeutics Division, Travenol Laboratories,
Los Angeles, CA 90039

S. R. Lowry, Nicolet Instrument Corporation, Madison, WI 53711

Hiroyuki Makino, Department of Polymer Technology, Tokyo Institute
of Technology, Ookayama, Tokyo, Japan

Sei-ichi Manabe, Textile Research Laboratory, Asahi Chemical
Industry Comp. Ltd., Takatsuki, Osaka, Japan

David P. Matzinger, Dynapol, 1454 Page Mill Road, Palo Alto,
CA 94304

P. Meares, Chemistry Department, University of Aberdeen, Scotland

Hiroki Narita, Department of Polymer Technology, Tokyo Institute
of Technology, Ookayama, Tokyo, Japan

Takashi Nohmi, Department of Polymer Technology, Tokyo Institute
of Technology, Ookayama, Tokyo, Japan

Y. Okamoto, Dept. of Chemistry and Polymer Res. Inst., Polytechnic
Inst. of N.Y., 333 Jay St., Brooklyn, NY 11201

W. P. Olson, Hyland Therapeutics Division, Travenol Laboratories,
Los Angeles, CA 90039

C. E. Pippenger, College of Physicians and Surgeons, Columbia University, New York, NY

W. Pusch, Max-Planck-Institut für Biophysik, Kennedy-Allee 70 6000 Frankfurt am Main, Germany

Eizo Sada, Chemical Engineering Department, Kyoto University, Kyoto 606, Japan

Masami Shiozawa, Chemical Engineering Department, Kyoto University, Kyoto 606, Japan

K. S. Spiegler, Department of Chemistry and Chemical Engineering Michigan Technological University, Houghton, MI 49931

Shmuel Sternberg, Millipore Corporation, Bedford, MA

H. Strathmann, Forschungsinstitut Berghof GmbH, D - 7400, Tübingen, Postfach 1523

Lori Tallman, Varian Instrument Group, 2700 Mitchell Drive, Walnut Creek, CA 94598

Mary K. Tripodi, Corporate Research and Development, Monsanto Company, St. Louis, MO 63166

Abraham Warshawsky, Department of Organic Chemistry, The Weizmann Institute of Science, Rehovot, Israel

C. Timothy Wehr, Varian Instrument Group, 2700 Mitchell Drive, Walnut Creek, CA 94598

Nurit Weiss, Polymer Department, Weizmann Institute of Science, Rehovot, Israel

Graham Yeadon, National Adhesives and Resins, Braunston, Daventry, Northants, England